THE PMM'S PROMPT PLAYBOOK

Mastering Generative AI for
B2B Marketing Success

David E. Sweenor

It's not the tech that's tiny, just the book!™

TinyTechMedia LLC

The PMM's Prompt Playbook:
Mastering Generative AI for B2B Marketing Success

by David E. Sweenor
Published By:
TinyTechMedia LLC

Editor: Peter Letzelter-Smith
Cover Designer: Josipa Ćaran Šafradin
Proofreader / Indexer: Peter Letzelter-Smith
Typesetter / Layout: Ravi Ramgati
February 2025: First Edition
Revision History for the First Edition
2025-02-18: First Release
ISBN: 979-8-9911299-3-0 (paperback)
ISBN: 979-8-9911299-4-7 (eBook)

www.TinyTechGuides.com

In Praise Of

Melissa Burroughs, Director of Product Marketing

The PMM's Prompt Playbook is a game-changer! David's book has completely transformed my approach to product marketing. Before reading this book, I'd only tried using AI for a few simple tasks, like automating repetitive processes or drafting basic content. I wasn't leveraging AI to help me craft PMM strategy—but now, thanks to David Sweenor's guidance, I'm doing more than I thought possible. I can go from high-level ideas to refined GTM strategy faster than ever before. The prompt workflow examples are incredibly useful, and the step-by-step instructions make it easy to apply gen AI to my daily tasks. This book is an invaluable resource for any PMM looking to unlock the power of AI in their daily workflows.

Dave Gerhardt, Founder, Exit Five

Forget the hype, this book delivers the no-BS, actionable advice that B2B marketers need to get started with prompts today. David breaks down complex concepts into digestible steps, providing clear examples and prompt workflows you can use immediately. Whether you're a seasoned PMM or just starting out, The PMM's Prompt Playbook will equip you with the skills and knowledge to thrive in the age of gen AI.

Rich Mendis, CMO and Expert Member of ISO/IEC JTC 1/SC 42 Artificial Intelligence

David Sweenor has written the essential guide to AI for product marketers. The PMM's Prompt Playbook is a treasure trove of practical advice and actionable strategies for leveraging AI across the entire marketing spectrum. This book will empower your PMM team to drive greater efficiency, improve decision-making, and achieve breakthrough results.

Dedication

To all of those who have been impacted by a RIF, stay positive and build your own destiny.

Prologue

TinyTechGuides are designed for practitioners, business leaders, and executives who never seem to have enough time to learn about the latest technology and marketing trends. These guides are meant to be read in an hour or two and focus on the application of technologies in a business, government, or educational setting.

After reading this guide, we hope that you'll have a better understanding of how generative AI is implemented in the real world—as well as a better idea of how to apply best practices in your business or organization.

Wherever possible, I try to share practical advice and lessons learned over my career so you can take this learning and transform it into action.

Remember, it's not the tech that's tiny, just the book!™

If you're interested in writing a TinyTechGuide, please visit www.tinytechguides.com.

Table of Contents

CHAPTER 3

SECTION II

CHAPTER 4

CHAPTER 5

CHAPTER 6

CHAPTER 7

CHAPTER 8

CHAPTER 9

SECTION III

CHAPTER 10

CHAPTER 11

CHAPTER 12

CHAPTER 13

CHAPTER 14

About this Book

This book is trisected into the following sections.

Section 1: Foundations. Chapters 1–3 introduce generative AI and key concepts that are a prerequisite for success.

Section 2: Core PMM Activities. Chapters 4–8 provide detailed workflows for core PMM activities like market research, positioning, messaging, product launches, sales enablement, and content creation.

Section 3: Related Disciplines and Applications. Chapters 9–12 build upon the foundations but provide less detailed workflows related to the myriad of special projects, applications, and related activities that are increasingly expected of PMMs.

This book is not traditional in the sense that it is expected to be read end-to-end like other TinyTechGuides. Section 1 is designed with that in mind. However, afterwards there are an increasing number of prompt workflows that can cut-and-pasted into your favorite generative AI service.

A digital copy of the prompts can be found at
www.tinytechguides.com.

Foundations

Introduction to AI for PMMs

I once had a business school professor who said, "Technological change is inevitable. These changes can chew up a generation of workers but productivity improves and, for the most part, people's lives also improve as a result of these changes." For some odd reason, that comment stuck with me throughout the years and seems more relevant today than ever.

Let's discuss the elephant in the room—AI is here and has fundamentally changed the nature of work for marketers. Some product marketing managers (PMMs) are nervous and slow to adopt generative AI, while others use it for just about everything. Will it take our jobs? Maybe.[1] But sitting idly by isn't the answer. AI is another tool in the toolbox, one that can make work more efficient.

Today's PMMs are in a tight spot. As a PMM, you're expected to be an expert in your product or solution, market, competitors, buyer personas, user personas, industries, partner ecosystems, and customers. You need to craft impactful positioning and messaging, create compelling demand-generation content, develop go-to-market strategies, and create sales enablement programs that drive results. And you needed to do it all yesterday, with a skeleton crew.

Generative AI tools like ChatGPT and Perplexity can mean working faster and smarter by automating many of the time-consuming tasks that PMMs face daily. Need to quickly research a new market or create a dozen variations of a value proposition targeted towards a specific persona? AI can help with that. Need to create a competitive matrix for a particular set of product capabilities? AI can help.

However, although AI can do many things, it can't replace your intuition and expertise.

In short, think of generative AI as a copilot and brainstorming buddy for B2B PMMs. It's not about replacing human creativity and expertise, but rather augmenting it. Seriously, do you really want to write another sales outreach cadence or landing page copy for the latest whitepaper you've spent the last month working on? What about the webinar that you just delivered?

By using AI to handle some of the repetitive drudgery and help augment research, analysis, and content creation, PMMs can focus on crafting compelling thought leadership content, building cross-functional relationships, and working with integrated campaigns, field marketing, and sales (revenue) enablement to drive awareness, conversions, pipelines, and revenue.

When I looked for a practical guide on using generative AI for B2B product marketing, I came up empty-handed. This book aims to fill that gap. It complements insights from TinyTechGuides' Modern B2B Marketing: A Practitioner's Guide for Marketing Excellence.[2] The difference? This TinyTechGuide provides a practicum on creating prompts using tools—readily available today—that will make your life easier.

This isn't a "you could use a prompt in this way" book. It's a hands-on guide based on my experience using prompts in my day-to-day workflows. All of the prompts and strategies discussed have been tested by me or my colleagues. Although I may change some of the actual details to protect the companies I've worked for, each prompt will capture the essence of the tasks and workflows. Near the book's end, things will start to get repetitive—and that's

OK. This means you are learning how best to structure prompts to get the most out of them.

I'll primarily use ChatGPT and Perplexity.ai for the examples to keep things simple. But you can use any of the tools out there. Obviously, there are nuances between these—the tech is rapidly evolving and new capabilities are always being added (almost daily). But, I hope that the strategies for thinking about how to use generative AI to accelerate your PMM workflows will endure.

Who Is This Book For?

The guide is meant for B2B PMMs looking to gain an edge with generative AI. Whether it's fleshing out personas, working on persona-based messaging, competitive analysis, planning product launches, prepping for the latest Gartner Magic Quadrant, or creating enablement materials, you'll find step-by-step workflows here.

This book is designed specifically for B2B product marketing managers who—by using the generative AI toolset—want to work smarter, not harder.

If you're a PMM looking to:
- Better understand your target market segments.
- Gain deeper insights into customer needs and preferences.
- Create targeted messaging that resonates.
- Research and develop rich, insightful personas in a fraction of the time.
- Scale your content creation efforts without sacrificing quality.
- Streamline your go-to-market strategy and execution.
- Plan better product launches.
- Tackle the never-ending stream of special projects with ease.
- Free up time for higher-level strategic thinking (this is more of a dream since most companies keep putting more things onto your plate without taking anything off).

Then this guide is for you.

Whether a seasoned PMM or new to the role, the prompts and strategies in this book will help you level up your product marketing game.

What This Book Is Not

Let's be clear about what this guide won't cover before exploring how you can use AI to simplify your PMM marketing life.

This book isn't about scaling AI adoption across an entire marketing department; while important, that's beyond its scope. I'm focused on helping individual PMMs become generative AI gurus who can drive greater business outcomes and impress the boss.

Also, this book isn't about, teaching you how to be a PMM. Most of that comes with experience. The Modern B2B Marketing book mentioned earlier is a good place to start if you need a primer.

If you're seeking a high-level overview of generative AI use cases, check out TinyTechGuides' Generative AI Business Applications: An Executive Guide with Real-Life Examples and Case Studies.[3] This TinyTechGuide, on the other hand, is hands-on with prompts and step-by-step examples.

This guide won't make you an AI expert overnight but it will provide practical know-how and confidence to start integrating AI into your daily work. Consider it a PMM quick-start for those struggling to keep up with the ever-increasing demands of a modern-day PMM.

Beginning in Chapter 4, practical exercises that you can run with your team are included. Of course, I'm always available to instruct a class if that's your fancy.

Tools of the Trade

Now, there are dozens of different services out there. From Google Gemini to Microsoft's Copilot, to Mistral AI's Le Chat and Anthropic's Claude, they have their quirks and nuances but all work basically the same. Let me share the key tools I use daily as a PMM:

ChatGPT

Probably the most popular tool is OpenAI's ChatGPT.

- **Best for**: Quick drafts, brainstorming, message testing.
- **Real use case**: I use it to draft initial positioning and messaging frameworks, along with blog and presentation outlines that I then refine.
- **Tip**: Works best with clear context and specific requests.

Perplexity

The Perplexity.ai-powered search engine is a powerful tool.
- **Best for**: Market research, competitive updates.
- **Real use case**: I ask it to analyze recent industry news and summarize key trends.
- **Tip**: Great at finding and citing current sources.
- **Bonus tip**: There's a great "Rewrite" function that allows you to select models like Claude, Grok, GPT, and Sonar (based on LlaMa 3.1 70B).

How can I use these tools?

As a PMM, you can use these tools in any number of ways but here is an example workflow to get the gears turning in your brain.
- Perplexity: Research competitor news.
- ChatGPT: Conduct in-depth analysis.

So, what tool should you use when? Perplexity is most useful when you need to have up-to-date current market news or competitive updates. ChatGPT is great for brainstorming, summarizing, and refining results.

> **Pro tip:**
>
> Use Perplexity for research and ChatGPT for content summarization and iterative refinement.

Getting Started

By now, you're hopefully convinced that generative AI can help you level up your PMM game. But where to begin? Here's a simple three-step process to get started:

1. **Identify the thing you hate doing most**: Start by identifying aspects of your work where AI could have the most significant impact. Maybe that's just those time-consuming tasks you hate doing the most, like adapting messaging to different personas, writing sales cadence emails, or creating yet another riff for the latest web page re-do. Focus on tasks that are time-consuming, repetitive, mentally exhausting, or that require a lot of research.

2. **You don't learn by watching**: Next, pick an AI tool and start experimenting. As of the writing of this book, most tools are around $20 per month. I'd recommend subscribing, though you can use the free versions as well. Begin with simple prompts and gradually increase complexity as you gain confidence and learn how the different models and services work. The key is to actively learn through hands-on experience.

3. **Track and measure results**: When integrating AI into PMM workflows, track results and write down the prompts used. How much time are you saving? Are you seeing improvements in the quality or quantity of the output? Use these insights to refine prompts and processes over time.

Pro tip:

Document and write down your prompts. Build up a personal prompt library that can be reused next time. Don't forget to share them with your peers; as they say, a rising tide lifts all boats.

Remember, the goal isn't to become an AI expert overnight. It's to gradually incorporate AI into your daily work in a way that delivers real value. So, start small, experiment often, write down prompts, and iterate to your ideal outputs.

Practical Advice and Next Steps

- **Use AI as a tool, not a brain replacement**: Consider AI as an accessory to improve productivity and creativity, not a replacement for your subject matter expertise.
- **Start small and experiment**: Begin by using AI for specific, manageable tasks like brainstorming and iterating content outlines. Gradually expand and refine uses as you become more comfortable.
- **Stay human**: Always review and refine AI-generated content to ensure accuracy, originality, and alignment with your brand voice and messaging.

Summary

- **Generative AI can help**: AI tools like ChatGPT and Perplexity.ai can help B2B PMMs accelerate research, content creation, content summarization, brainstorming, and planning.
- **Strategic and mindful use is essential**: To maximize benefits, PMMs should identify the time-consuming things they hate to do the most, experiment with Perplexity and ChatGPT to lessen that dynamic, and always write down their prompts as they go along.
- **Human expertise is still important**: While AI can automate tasks and make you more productive, human oversight, critical thinking, and creativity remain essential for ensuring quality, originality, and messaging alignment.

Chapter 1 References

[1] Sweenor, David. "Will AI Take My Job? Maybe." Medium. Last modified March 6, 2024. https://medium.com/@davidsweenor/will-ai-take-my-job-maybe-2c85ebf9e0a7.

[2] Sweenor, David, and Kalyan Ramanathan. Modern B2B Marketing: A Practitioner's Guide to Marketing Excellence. TinyTechGuides, 2024.

[3] Sweenor, David, and Yves Mulkers. Generative AI Business Applications: An Executive Guide with Real-Life Examples and Case Studies. TinyTechGuides, 2024.

CHAPTER 2

AI and Human Collaboration Framework

When it comes to using generative AI for product marketing, make sure you understand the roles that machines and humans should play. As mentioned earlier, think of AI as a brainstorming buddy and copilot. It can help with many tasks but needs clear direction and oversight from you, the subject matter expert (SME).

Understanding AI's Role

Think of large language models (LLMs) functioning something like spreadsheets. Spreadsheets take numerical inputs and apply formulas and calculations to produce mathematical results. Similarly, LLMs use formulas and calculations to process text inputs to generate text outputs that are statistically similar to the corpora they were trained on. And like spreadsheets, their usefulness depends entirely on the operator's skill and judgment. If you have bad data inputs within a spreadsheet—formulas or calculations—you'll get inaccurate results.

Here's how this translates to day-to-day work:

AI-led tasks

These are repetitive, time-intensive chores that AI can accelerate:
- Summarizing competitor updates from online sources.
- Drafting initial outlines or content variations tailored to specific personas or industries.
- Analyzing survey or customer review data to identify emerging trends.

Using AI for these tasks frees up your bandwidth for other priorities.

Human-led tasks

Tasks requiring judgment, strategic thinking, and deep domain expertise remain best handled by PMMs. These include:
- Developing and refining long-term GTM strategies.
- Creating and approving content for consistency with messaging frameworks.
- Managing key relationships with stakeholders and partners.
- Making business decisions based on organizational context.

PMM oversight ensures that outputs align with organizational goals and values.

Collaborative tasks

The most significant value often comes from combining AI's speed with human insight. For example:
- Using AI to draft a competitive analysis, followed by a PMM validating and tailoring the insights.
- Generating tons of message variations with AI, then testing and refining them collaboratively.
- Conducting market research to understand trends while the PMM interprets their implications.

When assigning tasks, begin with smaller AI-led responsibilities to evaluate outputs, then integrate AI into collaborative processes to maximize efficiency and creativity.

Think of it like this: Would you let a spreadsheet make investment decisions? No. But you'd use one to crunch the numbers. Apply the same logic to AI in your workflow.

Incorporating Generative AI into Workflows

Here's a workflow that exemplifies human-machine collaboration.

1. **Human: Write clear prompts**. As a PMM, it's your job to provide explicit instructions and organizational context for any AI-assisted task. Without ambiguity, ask specific questions and be direct regarding the required output format. Do you want five ideas or ten? Do you want the output in a table? If so, what columns do you want in the table? Better prompts yield better results.

2. **AI: Brainstorm initial content outlines**. Let the LLM do what it does best, quickly iterate and create a large number of ideas and adaptations based on your prompt. This is where the machine's speed and scale come into play. It's a useful copilot that can create countless versions and variations in a matter of seconds.

3. **Human: Validate outputs**. Review the AI-generated content with a keen eye. Is the output aligned with your messaging? Is it using language specific to the target persona? Are the technical claims, features, functions, and case studies real or made up? As the SME, it's up to you to recognize errors and inconsistencies.

4. **Human: Curate and refine**. Don't settle for the first output. Take the best parts of the AI-generated content and refine them further. Add your own insights, examples, and perspectives. Adapt the tone and style to match your unique point of view (POV) and your company's brand guidelines.

5. **AI: Iterate based on feedback**. If needed, feed the refined version back into the AI model and ask it, as an expert editor, to analyze the structure of the content. What

can be done to strengthen the content or make it more concise? This iterative process can help you get close to the final output, though you'll still have to edit and refine. But, a word of caution for the iterative process—you can keep putting the same content into ChatGPT and ask it to refine and edit; this is the fastest path to mediocrity and AI entropy.[1] As Kenny Rogers sang in "The Gambler": "You got to know when to hold 'em, know when to fold 'em."

By combining these endless adaptations and ideas from machines that have some of the creativity and expertise of humans, you can more efficiently create product marketing materials that are personalized and impactful.

Common Hazards to Avoid

When integrating generative AI into PMM workflows, there are a few common traps to watch out for:

Confabulations and lies

While AI outputs initially seem impressive, the technologies themselves are not truth seekers. As you experiment more, you'll start to notice their repetitiveness, and that they sometimes generate content that is inaccurate, outdated, and biased. Hallucination rates are all over the map. Some studies indicate that AI can hallucinate anywhere between 2.5 percent, 8.5 percent, more than 15 percent, or up to 69 to 88 percent of the time, depending on the domain and task.[2] Would you go to a doctor or get in an airplane with those kinds of accuracy rates?

The solve: Never trust and always verify

Always fact-check AI-generated content, especially when it comes to technical details, statistics, or claims about your product or those of competitors. When I say fact-check, go look at the original source, not a source that cites another source. With your prompt, the GenAI bot can be instructed to not add any facts that aren't present in the knowledge you provide directly (the quality control process outlined in the next section can be used). Remember, you're the expert and the final decision-maker.

A real-world GenAI lie

I was using generative AI to assist with a Gartner Magic Quadrant RFP. I decided to upload product documentation to our enterprise implementation of ChatGPT. When I asked it to help respond to a question, it outputted something that I knew was wrong. I asked it for the page number and specific quote that it cited. It dutifully shared the page number and specific quote. When I looked in the manual, it didn't exist.

The lesson? Always go look at the source, even if the AI convincingly provides you with a citation, quote, and page number.

Overlooking organizational context

Generative AI is trained on the world's data, but the models don't know anything about your organization besides what they've read on its web pages. Unless you tell it to behave in a certain way, AI output will not be specific to you or your company's tone, style, messaging, or brand guidelines.

The solve: Provide relevant context

The more context provided to the AI model, the more relevant and accurate its outputs (we'll discuss context and knowledge management more in Chapter 3). If you're using AI to generate content for multiple channels or personas, be sure to feed it your messaging framework, persona details, and brand guidelines. Are customer case studies, product details, and thought leadership survey data missing from the AI-generated content? The AI can only work with what you give it, so be sure to add your unique context.

Lacking specificity

Vague or overly broad prompts will yield equally vague and generic results. The more specific and targeted inputs, the more relevant and useful the outputs.

The solve: Use specific prompts

Avoid vague or open-ended prompts like "Write a blog post about our product." In fact, never ask it to blindly write blog posts or content about anything. Always start with an outline, iterate, then refine the outline. Ask it to create an initial high-level outline, ask it to pause after each section, and then ask it to query you with questions for feedback before moving on to the next section.

Take the time to craft clear, detailed prompts. You should exhibit the role or persona you'd like the LLM to adopt, the domain you're operating in, the input context, the task to achieve, and the output format you'd like. Do you want a downloadable CSV table, a narrative, or bullet points? Do you want the section headlines in title or sentence case? With or without emojis? Ask and you will receive. We'll discuss prompt design in Chapter 3.

> **Pro tip:**
>
> Did you know that if you're nice to LLMs they generate better content?[3] So, say "please" and "thank you" and treat the AI respectfully.

Disclosing sensitive information

AI tools are not inherently secure. Sharing sensitive data can expose your organization to privacy breaches or regulatory violations. Even the most popular platforms may not meet enterprise compliance standards.

The solve: Avoid inputting proprietary information into third-party systems

Use approved, secure tools and anonymize data where possible. Adhering to your company's privacy policies ensures safe and responsible use of AI.

The key is using AI thoughtfully and strategically, not blindly. By being aware of possible pitfalls and taking steps to mitigate them, you can become a rockstar, AI-powered PMM.

Quality Control Process

Since LLMs have a propensity to make sh** up, you'll need to create a quality control process to help monitor and systematically review the output. Below are some quality control checks you can build into your prompts to ensure messaging alignment, customer relevancy, technical accuracy, and business impact.

Messaging alignment

As previously mentioned, LLMs can create off-brand content. You can use LLMs to:
- Check adherence to positioning frameworks.
- Verify value proposition consistency.
- Maintain competitive differentiation.
- Follow approved message hierarchy.

Sample prompt: "Using our messaging framework {insert messaging framework}, analyze this content for alignment. Identify any messages that: 1) deviate from our core positioning, 2) miss key differentiators, or 3) don't match our target audience's needs."

Now, in my experience, it should be noted that messaging frameworks are either 1) contained on one page in a weird nested table structure, or 2) captured in a rather long and lengthy document. For whatever reason, there is no in-between. So, for the sample prompt above, at the {insert messaging framework} step, attach the document rather than inserting it in-line as illustrated above. Alternatively, use the XML tag style, which will be covered in Chapter 3.

Customer relevance

As a PMM, your primary goal is centered on customer relevancy. LLMs can be used to:

- Map to identify customer needs.
- Confirm use case applicability.
- Test competitive claims.
- Validate solution fit.

Sample prompt: "Based on our buyer personas {insert buyer personas}, evaluate this content for relevance. Flag any sections that: 1) don't address key pain points, 2) miss critical business challenges, or 3) fail to demonstrate clear value."

Technical accuracy

Remember the documentation hallucination mentioned earlier? To ensure technical accuracy, you can use generative AI to:

- Verify all technical claims against source documentation.
- Cross-reference product specifications.
- Check numerical claims and statistics.
- Validate citations and references.

Sample prompt: "Review this content for technical accuracy. Compare each technical claim to our product documentation. List any discrepancies or unsupported claims. Format the output as a table with columns for Claim, Source, and Validation Status."

Business impact

LLMs can even be used to help measure business impact. To that end, you can use AI to:

- Measure how messaging impacts sales cycles
- Track the effectiveness of competitive positioning
- Assess content usage by sales
- Monitor market category recognition

Sample prompt for a launch plan: "Review our Q4 product launch messaging. Show where it: 1) addresses our target market's top pain points, 2) strengthens our competitive position against {competitor names}, and 3) supports our revenue goals for enterprise deals over $500K. Format as a table with gaps and recommendations."

Sample prompt for sales enablement: "Analyze our sales battlecards for {product name}. Identify which competitive claims: 1) have the highest win rates, 2) resonate most with IT leaders, and 3) need updating based on recent market changes. Include specific deal examples."

Pro tip:

Create a validation checklist for your prompts. Define what "good" looks like before you start generating content. Also, be sure to save quality control prompts—they'll save you time on future reviews.

Practical Advice and Next Steps

- **Set clear goals and expectations**: Define objectives and desired outcomes before starting any AI-assisted task to guide prompts and effectively evaluate the results.
- **Provide context and specificity**: Give the AI model relevant background information and use specific, detailed prompts to get more targeted and useful outputs.
- **Maintain a human touch**: Remember that AI is a tool. Always review, edit, and refine AI-generated content to ensure quality, accuracy, and alignment with brand voice.

Summary

- **Understand AI's role**: Take a collaborative approach, with humans providing clear prompts, context, and refinement, while AI generates ideas, outlines, and early first drafts.

- **Avoid the pitfalls**: Don't be lazy and settle for mediocrity. Use messaging guides, customer case studies, and company-specific details as the source of truth for your brainstorming buddy.
- **Build quality control prompts**: Since AI makes stuff up, build quality control prompts to help minimize the risk of erroneous content.

Chapter 2 References

[1] Sweenor, David. "AI Entropy: The Vicious Circle of AI-Generated Content." Towards Data Science, July 14, 2023. https://medium.com/towards-data-science/ai-entropy-the-vicious-circle-of-ai-generated-content-8aad91a19d4f.

[2] Emslie, Karen. "LLM Hallucinations: A Bug or A Feature?" Communications of the ACM, May 23, 2024. https://cacm.acm.org/news/llm-hallucinations-a-bug-or-a-feature/.

[3] Yin, Ziqi, Hao Wang, Kaito Horio, Daisuke Kawahara, and Satoshi Sekine. "Should We Respect LLMs? A Cross-Lingual Study on the Influence of Prompt Politeness on LLM Performance." arXiv preprint arXiv:2402.14531 (2024).

CHAPTER 3

Prompt Engineering

As a PMM, I've found that getting the most out of LLMs starts with understanding what they actually do. Think of them like gigantic auto-complete text processors—you put in text, you get text back. Of course, they can create images, video, and audio but I'm focusing on text—what PMMs deal with most frequently. You can certainly use it to create an outline and images for a sales presentation, but I'm going to shy away from discussing AI presentation builders.

Pro tip:

Both Perplexity.ai and ChatGPT have web search capabilities. Use follow-up queries like, "Please cite the sources for this claim using a Chicago-style format and include the URLs." When you get the source, go look at the web page and verify its authenticity. I've had many sources come back with AI-generated drivel. You can even ask it to use "reputable sources," but again, use your eyeballs and actually read the sources to make sure they're in alignment with your worldview and content standards. I've spent many hours trying to track down bogus citations from the ground truth source and have come up with nada. Or, you find citations that are ten-plus years old.

Conversation Design

Getting good results begins with clear instructions. These are known as prompts. Fundamentally, prompts are the instruction manual for the LLM to follow. If you've ever put together IKEA-style furniture, it's quite analogous. They provide a detailed set of instructions that tell you exactly what to do, and in what sequence to do it. When putting together that new IKEA desk, if all goes well and you follow the instructions, you'll only have a few nuts and bolts left over.

Prompt design

Be sure to structure prompts with clear markers. I use XML-style tags to organize prompts for more complex requests. I'm not saying I do this every time since it's more natural to have a simple back-and-forth chat with the LLM but I do use XML-style tabs on my initial requests.

Here's my go-to format:

<role>, <context>, <task>, <format>, <tone>, <validation (optional)>

Here's an example of the XML-style prompt structure:

<role>

[Specify the role or perspective the LLM should adopt. Example: "Data scientist," "SEO Expert."]

</role>

<context>

[Provide background information, requirements, or constraints relevant to the task. Include details like the industry, goals, or audience.]

</context>

<task>

[Clearly define the specific request. Include any steps or sub-tasks needed.]

</task>

<format>
[Describe the desired format of the response: "Provide a bullet list," "Create a table," "Write in paragraphs."]
</format>
<tone>
[Define the desired tone: "Technical," "Conversational," "Formal," "Persuasive."]
</tone>
<validation>
(Optional) [Specify how the output will be evaluated for success: "Should include 3–5 actionable points," "Ensure all legal constraints are addressed," "Include only verifiable data."]
</validation>

Another structure is using hashtags.

Here's a prompt that I use to create outlines.

ROLE

You are an expert editor for [insert company].

CONTEXT

I am writing a thought leadership article for business leaders. I want to create a short blog post discussing the attached whitepaper.

TASK

Create a short blog outline based on the attachments and URLs.

FORMAT

As an industry consultant advising a CEO or other business leader.

TONE

Straightforward, professional

AUDIENCE

My readers are CIOs, CDAOs, CEOs, and practitioners in data, analytics, and AI. Tailor your post to target what this audience looks for in executive advice.

RESPONSE

Create a blog outline based on the attachments.

Prompt variables

Variables become super helpful for reusability and rapid iteration in prompting. For example, the prompt:

"Write a {length} description of {product_name} for {audience}"

Using the following inputs (aka, variables):

- {length=2 paragraphs}
- {product name=<your product name>}
- {audience=healthcare CIOs}

Becomes:

"Write a 2-paragraph description of <your product name> for healthcare CIOs"

Using variables in prompt templates, as shown in the example, is pretty helpful for PMMs. It allows for the creation of adaptable content that can be easily customized for different products, personas, lengths, and channels. We will be making heavy use of variables beginning in Chapter 4.

Additionally, PMMs can leverage shortcuts on their MacBooks (or equivalent PC functionality) to make these prompts even more reusable. By creating shortcuts for frequently used prompt variations or specific product descriptions, PMMs can quickly insert these elements into their templates with just a few keystrokes. This reduces the chance of errors and speeds things along.

Key prompt formatting tips:

- Most of the providers suggest using triple quotes """ for long-form content, which is also used in many of the Python code examples. Personally, I find it confusing and prefer the <bracket/> content </bracket> or # content # formats.
- Separate instructions with line breaks.
- For more complex requests, number sequential tasks: [1], [2], [3].
- Use {variable} for replaceable elements.

Role-Based Prompting

I find when writing prompts that giving the language model a specific role or persona dramatically improves the outputs. Here's what works.

Expert personas

Instead of: "Write about data security."

Try: "You are a CISO at a Fortune 500 financial services company. Explain your top three data security concerns for 2024 and how you're addressing them."

Domain assignments

<role>
Enterprise Data Architect
</role>
<task>
Review our new data catalog product and highlight integration challenges a typical customer might face.
</task>
<audience>
IT Implementation Teams
</audience>

Audience adaptation

For the same product update, I might use:
- **CIO persona**: Focus on strategic value and ROI.
- **Technical architect**: Emphasize implementation details.
- **End user**: Highlight usability improvements.

Real-world example:
<role>
Senior Data Analytics Manager at a retail company
</role>
<context>
You're evaluating analytics platforms for a company-wide deployment

</context>
<task>
Write a pros/cons analysis of implementing our solution, focusing on scalability and user adoption
</task>

The key is making the role specific enough to shape the perspective, but not so narrow that it limits the useful insights. Often, I'll test the same prompt with different roles to get differing perspectives on the same topic.

Model Memory and Context

Memory, in the context of LLMs, refers to the model's ability to retain and reuse information from previous interactions within a conversation. This is quite important for maintaining context, understanding references to earlier points, and generating coherent and relevant responses. Here's what I've learned about managing how generative AI remembers and forgets things during conversations:

Session memory limits

Think of it like being in a meeting room with a whiteboard—once it's full, you need to erase something to add new content. When the model hits its limit (it could be as little as 4,000 words with the addition of knowledge and custom instructions), older context starts dropping off.

Pro tip:

When I was younger, I once went to traffic court to fight a speeding ticket. Not having the inclination to hire a lawyer (it was a small ticket), I decided to get a few books and try and fight the citation on my own. When I got there, one of my books recommended to object during the police officer's testimony and state that the officer is reading from the citation. So, I did. The Judge sustained, stating that the

Officer needed to recall this from his own independent recollection.

As he continued, he kept reading from the ticket and I objected again. This time, it was rejected. The judge said, "The officer may need to refresh his memory quite frequently."

So, like the policeman who pulled me over, make sure you refresh the LLM memory quite frequently.

Example of context management:

<previous_context>

Key points from our earlier discussion:

- Target market: Financial services.
- Core features: Real-time analytics, compliance reporting.
- Main competitor: Databricks.

</previous_context>

<new_request>

Using these parameters, draft the next section of our pitch deck

</new_request>

Context windows

Think of a context window of the amount of text that can be put into the LLM's memory. To handle longer or more complex requests, try the following:

- Break work into smaller chunks.
- Summarize key points periodically.
- Start fresh conversations for new aspects.
- Save important outputs separately.

Context window example:

<summary>

From our previous exchanges, we defined:

- Product positioning: {positioning}
- Target personas: {personas}
- Key messages: {messages}

```
</summary>
<new_focus>
```
Now, let's develop the sales enablement cheat sheet based on these foundations. Please generate an outline for the cheat sheet.
```
</new_focus>
```

Data persistence

LLMs have the following types of memory:
- **Session memory (temporary):**
  - Lasts only during a current conversation.
  - Clears when you start a new chat.
  - Limited by context window.
- **Custom instructions (persistent):**
  - Stays across all new conversations.
  - Sets baseline behavior.
  - Helps maintain a consistent tone and style.
- **Knowledge base (no memory):**
  - Base knowledge from training.
  - Can't learn or update from our chats.
  - Doesn't remember past conversations.

An example of using custom instructions (set once in preferences):

"You are a PMM at a B2B software company. For [your company name], always:
- Write in clear, business-focused language.
- Include relevant industry standards.
- Format for skimmability.
- Consider enterprise security requirements."

One thing I've learned is that in long conversations, LLMs are forgetful. I keep track of important points and restate them when needed. Oftentimes, I'll start fresh with a new prompt that includes the key details from our previous exchange.

> **Pro tip:**
>
> Recently, I've started to notice that LLMs are not following instructions when processing documents as they once did, and, as mentioned above, they're quite forgetful. For example, if you ask it to create a LinkedIn, Twitter, and Bluesky social media post for a list of ten different articles, it may only process four of them. As the context windows get longer, it costs more to run lengthy prompts.[1] You'll also notice that many providers are now defaulting to "concise" responses. Is this a cost-cutting measure by the service provider? Perhaps.

Priming the Prompt

To provide the LLM with context, it's important to feed it content that it may not have access to or was not included in the training corpora. Here are a few sources commonly used to feed external information into generative AI models.

Website content

If you're not familiar with reader mode on popular web browsers, look it up—it'll become your new best friend. You can paste content from competitor websites using:

- Safari or Chrome Reader Mode to clean up formatting.
- Specific page sections (like product features).
- Multiple pages for comprehensive view.

Example: "Here's text from Competitor X's product page about their data catalog: {paste cleaned content}. Analyze their positioning and feature emphasis."

YouTube analysis

For YouTube, transcripts are a great external source. To use YouTube content:

- Get video transcript (use YouTube's CC feature).

- Clean up transcript formatting.
- Segment by topic if it's long.

Example: Here's a transcript from Competitor Y's product demo.

<Transcript/>

[paste transcript]

</Transcript>

<Task/>

Extract their key messaging points and technical claims.

</Task>

Documents

Internal and external documents are quite useful to inform and provide context to LLMs. These documents can include:

- PDF reports (analyst reports, whitepapers).
- CSV data (customer surveys, usage stats).
- Blog posts.
- Marketing emails.
- Data from Marketo form fills (business titles, companies, industries, etc.).

Compare messaging across these sources:

- Product blog post.
- Competitor's YouTube demo transcript.
- Latest Gartner report excerpts.

Pro tip:

- Break lengthy content into smaller, ~2000-word chunks.
- Include source context for each input.
- You can save PDFs as TXT files to reduce file size limits.
- Ask for specific comparisons or analysis.
- Use consistent formatting for each source type.

Here's a real example:
<input>
I've attached:
• Transcript from Competitor X's keynote.
• Their product page content.
• Their latest blog posts.
Analyze their new product positioning and target market focus
</input>
<transcript>
[pasted transcript]
</transcript>
[pasted page content]
[/product page>
[pasted blog content]
</blogs>

Practical Advice and Next Steps

- **Structure prompts clearly**: Use clear markers like content </instruction> or # content # to give the AI model clear instructions and parameters.
- **Start simple and build**: Begin with basic prompts and gradually add more detail and complexity as the conversation progresses.
- **Break down complex tasks**: For larger projects, use a step-by-step approach, breaking the task into smaller, more manageable chunks. This is known as Chain of Thought prompting.[2]

Summary

- **Effective prompting is key**: The quality of output from language models depends heavily on clear, specific, and well-structured prompts.

- **Iterative refinement improves results**: Building on initial responses through follow-up questions and adjustments leads to more useful and accurate outputs.
  - **Context management is crucial**: Maintaining context throughout a conversation and breaking down complex tasks into manageable chunks helps language models generate coherent and relevant content.

Chapter 3 References

[1] Lee, Timothy B. "Why AI Language Models Choke on Too Much Text." *Ars Technica*, December 20, 2024. https://arstechnica.com/ai/2024/12/why-ai-language-models-choke-on-too-much-text/.

[2] Wei, Jason, Xuezhi Wang, Dale Schuurmans, Maarten Bosma, Brian Ichter, Fei Xia, Ed Chi, Quoc Le, and Denny Zhou. "Chain-of-Thought Prompting Elicits Reasoning in Large Language Models." *arXiv preprint* arXiv:2201.11903 (2022). https://arxiv.org/abs/2201.11903.

Core PMM Activities

Now that we have the core PMM foundations out of the way, we will begin getting into workflows for each section. As we progress from Market Research to Positioning and Messaging to Content and Communications, the workflows will become more sophisticated and there will be less explanation within each workflow.

CHAPTER 4

Market Understanding

One of the core functions of a PMM is to ensure bringing external perspectives to their company. Sadly, this gets more challenging as companies decrease the headcount of PMM teams. For many, there is simply not enough time to understand the market in depth. Happily, generative AI can help speed up the market research process. However, this is no substitute for talking and interacting with prospects, customers, sales, solution engineers, and customer success managers; however, it is a helpful jumpstart for those of us already stretched thin.

As mentioned in Chapter 3, adding the appropriate context to generative AI models is paramount for success. The good news is that organizations have no shortage of material that can be used to inform the generative AI model. You can incorporate industry reports, surveys, customer reviews, social media, and other sources to help provide a fuller picture to the prompt.

For example, when I was researching new analytics capabilities before a launch, I examined a variety of sources:

- Checked what Gartner, Forrester, IDC, and other analysts were saying about analytics trends.

- Used materials from reputable sources like McKinsey, *MIT Sloan Management Review (MIT SMR)*, and the *Harvard Business Review (HBR)* for other perspectives.
- Looked at the messaging and blogs of our top five competitors.
- Reviewed G2Crowd, TrustRadius, and Gartner Peer Insights about gaps in product, services, or GTM approach.
- Analyzed selected Reddit, LinkedIn, and Medium articles and comments.

The goal is connecting the dots between these sources to spot opportunities and risks. Just remember—data without analysis is just numbers. Take time to find the "why" behind what you're seeing. What's the story your data is telling?

This chapter will guide you through a structured, step-by-step approach to AI-assisted market research by looking at trends, marketing sizing, competitive intelligence, segmentation, and user needs.

Trend Discovery

Okay, now let's shift gears a bit and talk about using AI to spot those all-important market trends. After all, as PMMs, we need to be ahead of the curve, right?

Here's where those prompt best practices come in. Instead of just asking something generic like, "What are the trends in AI?" you want to be specific.

For example, you could use a simple prompt like this in Perplexity AI:

<context>

I need to understand the key trends driving enterprise AI governance in the coming year.

</context>

<task>

Search the web and identify the top three trends in enterprise AI governance trends for 2025, with a focus on the impact to financial services companies.

</task>
<format>
Provide a bulleted list with a short explanation of each trend.
</format>
See how that's much more focused? You're giving the AI clear directions and constraints.

Pro tip:

Don't just rely on the AI's knowledge base. Upload relevant whitepapers or reports to give it even more context. This will help you get richer, more data-backed insights.

However, this is only scratching the surface. The variables mentioned in Chapter 3 can be used to create reusable prompts that you'll store in your favorite prompt library.

We'll walk through the approach in the first two prompts; subsequent ones will follow a similar path.

To get started, you need to define the variables:

Inputs (variables):

- {industry=Insert your target industry, e.g., financial technology for SMBs.}
- {focus=Insert the specific focus area, e.g., emerging payment solutions or AI-driven compliance tools.}

Then, iteratively, paste these into ChatGPT:

Define the objective

You are a senior industry analyst with expertise in {industry}.
</role>
I need to identify and analyze key industry trends for {industry}, with a focus on {focus}. The goal is to uncover technological innovations, shifts in customer behavior, and regulatory changes that will shape the market over the next 12–18 months. </context>

Provide an analysis of current and emerging trends in the {industry}, focusing on the following areas:
- **Technological innovations**: Identify cutting-edge technologies and solutions transforming the industry.
- **Customer behavior**: Highlight shifts in customer expectations, demands, or behaviors.
- **Market dynamics**: Describe competitive shifts, emerging players, or consolidations in the market.
- **Regulatory changes**: Summarize recent or upcoming regulations impacting businesses in this industry.
- **Growth opportunities**: Identify untapped opportunities or emerging gaps in the market.
</task>
<format>
Provide the response in a structured bullet list format, with 2–3 concise points under each focus area.
</format>
<tone>
Professional, strategic, and forward-looking.
</tone>

Pro tip:

Notice the "**" in the prompt above? This is part of the vocabulary of Markdown, a lightweight markup language used for formatting text on the web. It was created by John Gruber in 2004 to create readable and portable documents. Markdown is easy to learn and write, and it can be used to create a variety of content, from simple text documents to complex web pages.[1]

Build context iteratively for deeper insights

You are an expert consultant helping executives understand emerging trends in the {industry}.

</role>
I need to develop a detailed understanding of the {industry} to inform strategic decisions. Start with foundational insights, then progressively dive into specific trends and implications.
</context>
Break down the analysis into the following steps:
- Step 1: Industry overview. Describe the current state of the {industry}, including its size, key players, and primary challenges.
- Step 2: Technology trends. What are the top 3 technological advancements driving transformation in {industry}? Include innovations like AI, cloud, or automation.
- Step 3: Customer trends. How are customer needs and behaviors evolving in response to industry changes? What recurring pain points or opportunities exist?
- Step 4: Regulatory landscape. Summarize recent or upcoming regulatory changes in {industry} and their implications for businesses.
- Step 5: Future outlook. Provide a forward-looking view: What are the key opportunities and risks for businesses over the next 3–5 years?
</task>
<format>
Organize responses into short paragraphs or bullet points under each step for clarity. </format>
<tone>
Analytical, strategic, and data-driven.
</tone>

Simulate stakeholder perspectives

You are an industry thought leader providing expert commentary on trends shaping the {industry}.
</role>

I am conducting a virtual interview to simulate the perspectives of key stakeholders in the {industry}. The goal is to gain unique insights into how these trends are perceived and addressed by businesses.

</context>

Answer the following questions as if you are a key industry stakeholder:

1. Vendor perspective:
   o What technological trends are most important to companies like yours in the {industry}?
   o How is your company preparing to capitalize on these trends?

2. Customer perspective:
   o As a decision-maker in this industry, what are your top challenges or priorities for the next 12 months?
   o How are you evaluating and adopting new solutions to address these challenges?

3. Competitor perspective:
   o How are competitors positioning themselves to adapt to the emerging trends in {industry}?
   o What strategies are they using to differentiate or maintain market leadership?

</task>

<format>

Provide responses as if speaking in an interview, with concise and realistic insights for each perspective.

</format>

<tone>

Insightful, conversational, and grounded in expertise.

</tone>

Pro tip:

You can simulate customer interviews and surveys with ChatGPT and Perplexity. In the above example, you can ask the AI to take on different roles.

Summarize trends and strategic recommendations

You are a senior strategist presenting key industry trends to an executive leadership team.
</role>
Summarize the most important trends in {industry} and provide actionable recommendations for businesses to capitalize on these insights.
</context>
- **Summarize the Top 5 trends**: List the most impactful trends in {industry}, with 1–2 sentences describing each.
- **Opportunities**: Highlight specific opportunities businesses can pursue to align with or respond to these trends.
- **Risks**: Identify potential risks or challenges that companies should monitor.
- **Recommendations**: Provide 3 actionable strategies for businesses to stay competitive, innovate, or address the identified trends.
</task>
<format>
Structure the response as follows:
- **Top trends**: Bullet list with brief explanations.
- **Opportunities**: Short bullet list of opportunities.
- **Risks**: Bullet list of key challenges or risks.

- **Recommendations**: Numbered list of actionable strategies.

</format>
<tone>
Executive-level, concise, and actionable.
</tone>

So, for the trend discovery we:
- Defined the objective
- Built the context iteratively
- Simulated stakeholder responses
- Summarized the trends

TAM Sizing

Next, let's turn to total addressable market sizing. Many PMMs are already working for an organization re-targeting specific markets and don't create TAMs; however, we include this for completeness.

Input (variables):
- {industry=Insert your target industry, e.g., enterprise data analytics.}
- {solution=Insert the type of solution, e.g., data observability tools.}
- {audience=Insert target audience, e.g., enterprise companies or mid-market businesses.}

Define the TAM objective

CONTEXT
I need to calculate the Total Addressable Market (TAM) for {solution} within the {industry}. The target audience is {audience}, and the calculation should include both current demand and projected market growth over the next 5 years.

TASK
Provide a TAM calculation for {solution} in the {industry} using the following steps:
- Step 1: Estimate the total number of target organizations

in the {audience} segment, including a breakdown by size, geography, or vertical (if applicable).

- Step 2: Define the average annual revenue opportunity (ARR) or spend per organization for solutions like {solution}.
- Step 3: Calculate the TAM by multiplying the total number of target organizations by the average revenue opportunity per organization.
- Step 4: Provide a concise explanation of the assumptions used (e.g., market penetration rates, pricing estimates, or adoption trends).

FORMAT

Provide the response as follows:

- Step 1: Total number of target organizations.
- Step 2: Average revenue opportunity per organization.
- Step 3: TAM calculation (number of organizations x revenue opportunity).
- Step 4: Assumptions and rationale used in the TAM calculation.

TONE

Professional, data-driven, and concise.

Build context iteratively for accuracy

ROLE

You are a strategy consultant with expertise in {industry} market sizing.

CONTEXT

I am validating TAM estimates for {solution} within the {industry}. I want to ensure the calculation accounts for realistic market conditions and is aligned with current trends.

TASK

- Step 1: Target market definition. Estimate the total number of target organizations in the {audience} segment. Break it down by:
  - Organization size (e.g., SMBs, mid-market, enterprises).
  - Geographies or industry verticals (if applicable).

- Step 2: Revenue opportunity. Provide an estimate of the average annual spend (ARR) for {solution} per organization, adjusted for company size or vertical.
- Step 3: Market growth rate. Project the compound annual growth rate (CAGR) for {solution} within the {industry} over the next 5 years, citing key drivers like adoption trends, technological advancements, or increasing demand.
- Step 4: Addressable market calculation. Multiply the target organization count (Step 1) by the average revenue per organization (Step 2) to calculate TAM. Adjust for projected growth based on the CAGR.

FORMAT

Provide the response in a structured format:
- Step 1: Target organizations breakdown.
- Step 2: Annual revenue per organization.
- Step 3: Market growth rate (CAGR).
- Step 4: Final TAM calculation and key assumptions.

TONE

Structured, analytical, and supported by evidence.

Validate with competitive context

ROLE

You are an industry analyst comparing TAM opportunities for {solution} across competing vendors in {industry}.

CONTEXT

I need to validate the TAM estimate by benchmarking it against competitors' market positions and industry growth trends.

TASK

Provide an analysis that includes:
- Comparison of the TAM for {solution} against similar solutions in the {industry}.
- Key market drivers or trends that may increase TAM over the next 3–5 years.
- How competitors are addressing or projecting the same market opportunity.

FORMAT
Summarize the findings in bullet points, with comparisons across:
- Competitors.
- Key trends influencing TAM.
- Risks or opportunities that could adjust TAM estimates.

TONE
Strategic, insightful, and comparative.

Competitive Landscape Mapping

Input (variables):
- {industry=Insert your target industry, e.g., enterprise data analytics.}
- {solution=Insert the type of solution, e.g., data observability tools.}
- {competitors=Insert known competitors, e.g., Vendor A, Vendor B, Vendor C.}

Define the objective

You are a competitive intelligence analyst with deep expertise in {industry} and {solution}.
</role>

I am conducting a competitive landscape analysis for {solution} within the {industry}. I need to identify key competitors, their strengths and weaknesses, and how they position themselves in the market. The goal is to uncover opportunities for differentiation and strategic positioning.
</context>

Provide a detailed competitive landscape analysis for the {solution} category in the {industry}. Include the following:
- **Competitor identification**: List 3–5 key competitors, including emerging players.
- **Product features**: Compare the core features and capabilities of these competitors.

- **Market positioning**: Summarize how each competitor positions itself (e.g., cost leader, premium solution, innovation-driven).
- **Strengths and weaknesses**: Highlight the top strengths and weaknesses for each competitor.
- **Gaps and opportunities**: Identify areas where competitors are falling short and suggest opportunities for differentiation.

</task>
<format>
Provide the response in a structured table format:

Competitor Name	Key Features	Positioning	Strengths	Weaknesses
Vendor A	Feature 1, 2, 3	Cost leader	Strong integration options	Limited scalability
Vendor B	Feature 1, 2, 3	Premium solution	Excellent customer support	Higher price point

Include a brief summary of gaps and opportunities after the table.
</format>
<tone>
Professional, analytical, and concise.
</tone>

Build context iteratively for depth

You are a strategy consultant conducting a deep-dive analysis of {solution} in the {industry}.
</role>
I need to develop a more detailed understanding of each competitor's strengths, weaknesses, and market strategies to refine our go-to-market positioning.

</context>

Break down the competitive landscape as follows:

- **Feature comparison**: List the unique and overlapping features of the top competitors. Highlight any differentiators or gaps.
- **Customer sentiment**: Summarize customer reviews or feedback from sources like G2, TrustRadius, or Reddit. Focus on common themes in positive and negative feedback.
- **Pricing models**: Outline how competitors price their solutions (e.g., subscription tiers, usage-based pricing, enterprise licensing).
- **Innovation trends**: Identify any emerging innovations or strategies competitors are using to differentiate themselves (e.g., AI integration, faster deployment times, custom solutions).

</task>

<format>

Provide the response in bullet points under each category. Use short, data-driven insights to describe each competitor's approach.

</format>

<tone>

Insightful, data-backed, and actionable.

</tone>

Simulate competitor perspectives

You are a senior executive at a competing company in the {industry} that offers {solution}.

</role>

I am looking to understand how competitors view their positioning and strategy relative to us. The goal is to simulate competitor perspectives to uncover potential blind spots or opportunities.

</context>

<task/>

Answer as if you are a senior executive at one of the key competitors.

- How do you position your product in the market relative to others?
- What do you see as your biggest strengths and differentiators compared to the competition?
- What challenges or weaknesses do you face in capturing market share?
- How do you perceive {our product/solution} as a competitor? What areas do you see as threats or opportunities for improvement?

</task>

<format>

Provide responses in short paragraphs for each question, adopting the tone of a confident industry executive.

</format>

<tone>

Strategic, competitive, and forward-looking.

</tone>

Summarize insights and recommendations

<role/>

You are a senior strategist tasked with summarizing competitive insights for executive stakeholders.

</role>

<context/>

I need a clear summary of the competitive landscape for {solution} in {industry}. This summary should identify key market trends, areas for differentiation, and strategic recommendations.

</context>

<task/>

Summarize the competitive landscape analysis as follows:

- Key competitors and their positioning.

- Major strengths and weaknesses across the competitive set.
- Gaps in the market that present opportunities for differentiation.
- Actionable recommendations for product positioning, messaging, or feature development to gain a competitive edge.

</task>

<format>

Structure the summary in a prioritized list with clear, actionable recommendations. Include a short bullet summary of key findings under each section.

</format>

<tone>

Executive-level, concise, and results-oriented.

</tone>

Market Category Definition

When I joined a previous company, we were stuck between several market labels—analytics, data prep, data science, and AI. None quite fit.

So, in retrospect, I decided to see if generative AI could help. Here's what worked: Ask AI to analyze how competitors describe their categories, then look for gaps or opportunities to carve out our own space. But the real insight comes from studying how customers describe their problems to their bosses when asking for budget decisions.

Quick example: Instead of calling ourselves a "data quality platform" (yawn), we noticed customers talking about "making data trustworthy for AI." This simple shift in category definition opened up better conversations with prospects.

I use AI as a sounding board with prompts like:

- "What's missing from these existing category definitions?"
- "How would a CIO explain this type of solution to their CEO?"
- "What search terms do buyers use when looking for solutions like ours?"

The key is avoiding category buzzwords that make eyes glaze over. Stay grounded in customer language. If you can't explain your category to your parents, it needs work.

One caution: Don't let AI push you toward trendy but empty category labels. Test potential categories with actual customers. Their reaction tells you everything.

For the workflow:

Inputs (variables):

- {industry=Insert your target industry, e.g., enterprise cloud infrastructure.}
- {solution=Insert the solution type, e.g., data observability tools.}
- {category=Insert the market category, e.g., cloud data observability.}

Define the market category objective

You are an experienced Gartner analyst specializing in {industry} and {solution}.

</role>

I need to clearly define and position the market category for {solution} within the {industry}. The goal is to explain the purpose of the category, its unique characteristics, and its differentiation from adjacent markets to inform messaging and positioning strategies.

</context>

Define the {category} market as follows:

- **Category definition**: Provide a clear and concise definition of the {category}.
- **Core problems addressed**: Describe the top 3–5 business problems or challenges that this category solves.
- **Key features/capabilities**: List the critical features or capabilities that solutions in this category typically include.

- **Audience**: Identify the primary roles, industries, and business sizes that benefit from solutions in this category.
- **Differentiation**: Explain how this category differs from or overlaps with adjacent markets.

</task>
<format>
Organize the response in bullet points under each section for clarity. </format> <tone>
Professional, concise, and authoritative.
</tone>

Build the market category in layers

You are a senior strategy consultant helping define the competitive positioning for {solution} within the {category} market.
</role>
I need a deeper understanding of the category to identify opportunities for differentiation and growth. Build this analysis progressively, starting with foundational insights and leading to a competitive landscape assessment.
</context>
Analyze the {category} market across the following dimensions:
- Historical context:
  o When and why did this market category emerge?
  o What changes in technology, customer behavior, or business needs led to its creation?
- Current state:
  o What is the current size and growth rate of the {category} market?
  o Who are the major players offering solutions in this category?
- Market drivers:
  o Identify the macro (e.g., economic shifts, regulations)

and micro (e.g., customer needs, tech advancements) factors driving the growth of the {category} market.

- Adjacent categories:
  o What other market categories overlap or compete with {category}?
  o How is {category} unique or complementary to these adjacent markets?
- Future outlook:
  o How is the {category} expected to evolve over the next 3–5 years?
  o What opportunities or risks should businesses in this category be aware of?

</task>
<format>
Organize responses into sections with subheadings: "Historical Context," "Current State," "Market Drivers," "Adjacent Categories," and "Future Outlook." Use bullet points for clarity.
</format>
<tone>
Analytical, forward-looking, and strategic.
</tone>

Simulate stakeholder perspectives

You are a key stakeholder providing insights into the emerging {category} market from a specific perspective: vendor, customer, or competitor.
</role>
I am conducting virtual interviews to uncover insights about the {category} market. Simulate perspectives from three types of stakeholders: vendors, customers, and competitors.
</context>

Provide responses to the following questions from each perspective:

- **Vendor perspective**:
  - o How do you position your product within the {category} market?
  - o What unique value does your solution offer compared to competitors?
- **Customer perspective**:
  - o What problems led you to explore solutions in the {category} market?
  - o What features or capabilities are most important to you when evaluating solutions?
- **Competitor perspective**:
  - o How do competitors differentiate their offerings in the {category} market?
  - o What gaps or shortcomings still exist that represent opportunities for innovation?

</task>
<format>
Provide responses in short paragraphs for each perspective: "Vendor," "Customer," and "Competitor." Focus on realistic insights and actionable details.
</format>
<tone>
Insightful, conversational, and stakeholder-specific.
</tone>

Identify opportunities and differentiation

You are a senior product strategist identifying opportunities for differentiation within the {category} market.
</role>
I need to uncover opportunities to position {solution} as a market leader in the {category}. Focus on gaps in existing

solutions, unmet customer needs, and emerging trends that align with our capabilities.
</context>

Analyze the {category} market to identify opportunities for differentiation:

- **Gaps in current solutions**: What unmet needs or recurring customer pain points exist in this category?
- **Emerging trends**: Highlight trends (e.g., AI, automation, compliance) that can create opportunities for innovation.
- **Unique capabilities**: Based on typical solutions in this category, what features or capabilities could set us apart?
- **Targeted positioning**: Recommend specific messaging or positioning strategies to differentiate {solution} from competitors.

</task>
<format>

Structure the response in bullet points under each focus area. Provide actionable recommendations for each opportunity identified.
</format>
<tone>

Strategic, actionable, and market-focused.
</tone>

Summarize findings and recommendations

You are an industry consultant summarizing insights for executive leadership.
</role>

Summarize the key findings and strategic recommendations for defining and positioning {solution} in the {category} market.
</context>

Summarize the analysis as follows:

- **Market category definition**: A concise explanation of the {category} and the problems it solves.
- **Top trends and drivers**: Highlight the key trends, drivers, and emerging opportunities shaping the market.
- **Competitive landscape**: Provide a high-level overview of competitors and their positioning.
- **Gaps and opportunities**: Identify the most significant gaps in current solutions and customer needs.
- **Strategic recommendations**: Offer 3 actionable strategies to position {solution} as a leader in the {category} market.

</task>
<format>
Provide the summary in numbered sections with concise bullet points for each finding and recommendation.
</format>
<tone>
Executive-level, results-oriented, and concise.
</tone>

Audience Segmentation Strategies

Market segmentation is a classic PMM skill. Usually, people think about defining the total addressable market (TAM), discussed my Modern B2B Marketing TinyTechGuide and the previous section. However, for many B2B PMMs at enterprise software companies, the reality is that the company you are working for has existing products and a target go-to-market that has been determined. So, for the purposes of this section, we will focus on user personas and segments.

As a PMM who's handled multiple product launches, I've learned that effective segmentation is about more than just dividing your market into neat categories. It's about understanding who really needs your product and why. However, don't go overboard on the number of segments. Most companies I've worked for only have the capacity to target two to three segments at any given time.

Let me walk you through my approach, I start by looking at how different groups use our product. For example, when working on our data analytics platform, we noticed IT leaders, finance executives, and operations managers each had distinct pain points. IT focused on integration and security, finance cared about cost reduction, and operations wanted better reporting capabilities.

Pro tip:

Upload your customer data (like whitepaper downloads or webinar registrations) into ChatGPT and ask it to identify patterns in titles, industries, and company sizes. I recently did this with our whitepaper data and discovered an unexpected segment of analytics professionals who were downloading our content—a group we hadn't actively targeted before.

What's working well is combining traditional demographic data with behavioral insights. For instance, we track which features different user groups actually use, not just what they say they want. This helps create more accurate, actionable segments.

Define variables:

- {industry=Insert your target industry, e.g., enterprise software for financial services.}
- {solution=Insert your product/solution, e.g., cloud-based data observability platform.}
- {audience=Insert your target audience, e.g., IT managers, compliance officers, or mid-market enterprises.}

Define segmentation objectives

You are a market research expert specializing in {industry} and {solution}.

\</role>

\

I need to develop audience segmentation strategies for {solution} to optimize targeting, messaging, and product positioning. The target audience includes {audience}.

\</context>

\

Define the objectives for this segmentation exercise:

- What are the primary goals for segmentation? (e.g., campaign targeting, feature prioritization, go-to-market alignment).
- What insights are most critical to achieving these goals?
- Provide a suggested approach for developing segmentation strategies.

\</task>

\<format>

Provide a clear and concise bullet list outlining segmentation goals, critical insights, and a suggested approach.

\</format>

\<tone>

Professional and focused.

\</tone>

Identify relevant segmentation dimensions

\

You are a strategy consultant helping define key segmentation dimensions for {solution} in the {industry}.

\</role>

\

I need to segment my audience into actionable groups that reflect their needs, behaviors, and priorities. Possible dimensions include demographics, firmographics, behavioral data, and pain points.

\</context>

Identify and list relevant segmentation dimensions for {solution}, including:

- **Demographic/firmographic**: Company size, industry vertical, location, revenue.
- **Behavioral**: Usage frequency, engagement levels, feature adoption patterns.
- **Needs-based**: Core pain points, challenges, and goals specific to {audience}.

Provide examples of how these dimensions can be applied to create meaningful audience segments.

</task>

<format>

Organize the response into three categories:

1. Demographic/Firmographic.
2. Behavioral.
3. Needs-Based: Include examples for each category.

</format>

<tone>

Clear and actionable.

</tone>

Build audience segmentation framework

You are an expert analyst creating an audience segmentation framework for {solution} in {industry}.

</role>

Based on the identified segmentation dimensions, I need to create a structured framework that organizes my audience into clear, actionable segments.

</context>

Develop an audience segmentation framework that includes the following:

- Segment names that reflect their unique characteristics (e.g., "Compliance-Focused Enterprises").
- Key attributes for each segment (e.g., size, behavior, needs).
- A brief description of each segment's primary challenges and opportunities.

</task>
<format>
Provide the response in a table format:

Segment Name	Attributes	Challenges	Opportunities
Example Segment 1	Industry: Financial Services	Managing compliance at scale	Automate compliance workflows
Example Segment 2	Company size: Mid-Market	Limited technical resources	Simplified onboarding tools

</format>
<tone>
Structured and concise.
</tone>

Identify pain points and needs

You are a customer research expert analyzing pain points and needs for {audience} in {industry}.
</role>
I need to understand the most critical challenges and needs of each audience segment to refine messaging, positioning, and product development.
</context>
For each identified segment, outline the top 3 pain points and their corresponding needs.

Example format:
- **Segment name**: [Insert segment name]
- **Pain point**: [What challenge does this group face?]
- **Need**: [What does this group require to address the pain point?]

</task>

<format> Provide the response as a bullet list, grouped by segment. </format> <tone>

Insightful and focused on customer needs.

</tone>

Prioritize key segments

You are a market strategist prioritizing audience segments for {solution} in {industry}. </role>

I need to prioritize the audience segments based on factors like revenue potential, ease of adoption, and strategic alignment.

</context>

Prioritize the top 3 audience segments by:
- Revenue potential.
- Ease of adoption (e.g., low onboarding barriers).
- Strategic fit with {solution}.

Provide a ranked list of segments, along with a short rationale for each.

</task>

<format>

List the prioritized segments in ranked order:
1. **Segment name**: [Short rationale].
2. **Segment name**: [Short rationale].
3. **Segment name**: [Short rationale].

</format>

<tone>

Strategic and results-driven.

</tone>

Align segments to tailored messaging

<role/>

You are a product marketer crafting tailored messaging for each audience segment. </role>

<context/>

I need to align audience pain points and needs to messaging strategies that emphasize {solution}'s value proposition. </context>

<task/>

For each segment, create tailored messaging that includes:

- **Segment name**
- **Core pain point**: The top challenge this segment faces.
- **Value proposition**: How {solution} solves the pain point.
- **Supporting message**: A specific feature, benefit, or outcome tied to the solution.

</task>

<format/>

Provide responses in the following format for each segment:

- Segment: [Segment name]
  - o Core pain point: [Challenge]
  - o Value proposition: [How {solution} solves it].
  - o Supporting message: [Feature or benefit that reinforces the value].

</format>

<tone>

Clear, customer-centric, and persuasive.

</tone>

Validate and iterate

<role/>

You are a customer insights analyst validating audience segmentation and messaging.

</role>

I need to ensure my audience segments and messaging align with real-world customer feedback and sales insights.

</context>

Review the following audience segments and messaging: [Insert audience segments and messaging here].

- Identify any gaps or inconsistencies based on customer interviews, feedback, or sales insights.
- Suggest improvements to ensure accuracy and relevance.

</task>

<format> Provide a short summary of feedback gaps and 2–3 suggestions for refinement.

</format>

<tone>

Constructive, analytical, and precise.

</tone>

Competitive Intelligence

Alright, now that we've gathered all this intel, let's talk about how to actually use it to analyze our competitors. AI can be a huge help here. It can quickly sift through mountains of information and pinpoint key takeaways.

Think about it: instead of manually comparing your product to competitors across dozens of features, you can use AI to do the heavy lifting. But you need to guide it with a well-crafted prompt.

Define the inputs:

- {industry=Insert your target industry, e.g., data observability for enterprise SaaS.}
- {solution=Insert your product/solution, e.g., cloud-based data quality platform.}
- {competitors=Insert key competitors, e.g., Vendor A, Vendor B, and Vendor C.}

Define the objective

You are a competitive intelligence analyst specializing in {industry} with deep expertise in market research.

</role>

I need to conduct competitive intelligence on {competitors} to better understand their positioning, strengths, and weaknesses. The goal is to identify opportunities for differentiation and inform our product strategy and messaging for {solution}.

</context>

Provide a competitive analysis for {competitors} with the following focus:

- **Company overview**: Key details (e.g., founding year, company size, funding).
- **Product capabilities**: Core features, capabilities, and differentiators.
- **Positioning**: How each competitor positions their solution in the market.
- **Strengths**: Areas where they excel (e.g., integrations, support, pricing).
- **Weaknesses**: Gaps, shortcomings, or customer pain points.

</task>

<format>

Provide the response in a comparison table with the following columns:

Competitor	Overview	Core Features	Positioning	Strengths	Weaknesses
Vendor A	[Details]	[Features]	[Market focus]	[Strengths summary]	[Weaknesses summary]
Vendor B	[Details]	[Features]	[Market focus]	[Strengths summary]	[Weaknesses summary]

</format>

<tone>
Structured, analytical, and focused on actionable insights.
</tone>

Analyze competitor positioning and messaging

You are a market strategist specializing in competitor positioning and messaging within {industry}.
</role>

I need to understand how {competitors} position themselves in the market, including their value propositions, target audience, and messaging themes. This analysis will help refine the positioning of {solution}.
</context>

Analyze each competitor's positioning and messaging:
- What is their **core value proposition**?
- Who is their **target audience** (e.g., SMBs, enterprises, specific industries)?
- What key messaging themes or claims do they emphasize in their marketing materials?
- How does this messaging differentiate them (or fail to differentiate them) from other competitors?
</task>

<format>
Provide the response in bullet points for each competitor:
- Vendor A:
    - Value proposition: [Short description].
    - Target audience: [Description].
    - Messaging themes: [Key claims or themes].
    - Differentiation: [Strengths or gaps in their messaging].
- Vendor B:
    - Value proposition: [Short description].
    - Target audience: [Description].
    - Messaging themes: [Key claims or themes].

o Differentiation: [Strengths or gaps in their messaging].

</format>

<tone>

Analytical, focused, and messaging-oriented.

</tone>

Identify product and feature gaps

You are a competitive product analyst assessing feature gaps and opportunities in {industry}.

</role>

I need to identify gaps or opportunities in competitor solutions compared to {solution}. Focus on product capabilities, usability, and areas for innovation.

</context>

Compare {solution} to {competitors} and identify:

- **Feature gaps**: Capabilities competitors lack but customers value.
- **Usability gaps**: Complex or manual workflows in competitor products.
- **Innovation opportunities**: Features or capabilities that could set {solution} apart.

</task>

<format>

Provide the response as follows:

1. Feature gaps:
   o Competitor 1: [Feature missing].
   o Competitor 2: [Feature missing].
2. Usability gaps:
   o Competitor 1: [Description of complexity or manual process].
   o Competitor 2: [Description of complexity or manual process].

3. Innovation opportunities:
   - o Opportunity 1: [Potential feature or capability to stand out].
   - o Opportunity 2: [Another feature or enhancement].

</format>

<tone>

Clear, opportunity-driven, and actionable.

</tone>

Simulate competitor perspectives

You are a senior product manager at {competitor}.

</role>

I am simulating competitor perspectives to understand their strategy, priorities, and perceived weaknesses in the market. This will help us anticipate their moves and strengthen our positioning for {solution}.

</context>

Answer the following questions as if you are a senior leader at {competitor}:

- How do you position your product relative to others in the market?
- What are your product's biggest strengths and differentiators?
- What are your biggest weaknesses or challenges in winning deals?
- How do you perceive {solution} as a competitor?

</task>

<format>

Provide realistic, role-played responses in short paragraphs.

</format>

<tone>

Confident, strategic, and competitor-focused.

</tone>

Prioritize competitive insights

<role/>

You are a strategy consultant prioritizing competitive intelligence insights to inform product and marketing strategies for {solution}.

</role>

<context/>

I need to prioritize the competitive intelligence findings to identify the most impactful opportunities for differentiation and messaging.

</context>

<task/>

Prioritize the key findings based on:

- **Market impact**: How significant is this finding in influencing customer decisions?
- **Competitive threat**: How strong is the competitor's positioning in this area?
- **Strategic opportunity**: How well can {solution} address this area to differentiate?

</task>

<format>

Provide the response in a table format:

Finding	Market Impact	Competitive Threat	Strategic Opportunity	Priority
Competitor A: Pricing Model	High	Moderate	Strong	High
Competitor B: Feature Gap	Medium	High	Strong	Medium

</format>

<tone>

Strategic, structured, and action-oriented.

</tone>

Summarize competitive insights and recommendations

You are a competitive intelligence strategist presenting findings to product and marketing leadership.

</role>

Summarize the competitive intelligence insights, highlighting opportunities for differentiation, product improvements, and messaging strategies for {solution}.

</context>

Provide a concise summary of findings and actionable recommendations:

- Key competitive trends and insights.
- Areas where competitors excel or fall short.
- Opportunities for {solution} to differentiate or innovate.
- Strategic recommendations for positioning, product development, or GTM strategy. </task>

<format>

Provide the response in the following structure:

1. Key trends and insights:
   o Insight 1: [Description].
   o Insight 2: [Description].
2. Competitive strengths and weaknesses:
   o Strengths: [Key areas competitors excel].
   o Weaknesses: [Gaps in their products or positioning].
3. Opportunities for differentiation:
   o Opportunity 1: [Brief description].
   o Opportunity 2: [Brief description].
4. Strategic recommendations:
   o Recommendation 1: [Actionable suggestion for positioning].
   o Recommendation 2: [Actionable suggestion for product development].

</format>

```
<tone>
```
Executive-level, concise, and results-focused.
```
</tone>
```

Why this workflow works

- **Multidimensional**: Analyzes competitors from product, positioning, and strategic perspectives.
- **Action-oriented**: Focuses on identifying opportunities for differentiation and innovation.
- **Structured outputs**: Ensures findings are clear, actionable, and presentation-ready.
- **Simulated perspectives**: Adds depth by anticipating competitor strategies and weaknesses.

Practice Exercises

- **Conduct a trend discovery exercise**: Define the target industry and focus area, then step through the provided prompts to analyze current and emerging trends. Simulate stakeholder perspectives and summarize findings and recommendations. Share your insights with the team and discuss how the analysis can inform your product strategy or competitive positioning.

- **Practice defining a market category**: Specify the target industry, solution type, and potential category name for a product or a hypothetical one. Follow the prompts to define the category objective, build the category definition in layers, simulate stakeholder perspectives, identify opportunities for differentiation, and summarize findings and recommendations. Consider how this exercise can help you carve out a unique space in the market and your guide messaging strategy.

- **Conduct a competitive intelligence analysis**: Define the target industry, your solution, and the competitors to be analyzed. Step through the prompts to gather key details about each competitor, analyze their positioning and messaging, identify product or feature gaps, simulate competitor perspectives, and prioritize the insights that are uncovered. Summarize findings and recommendations in a presentation for your product and marketing teams, highlighting opportunities to differentiate your solution and strengthen your market position.

Summary

- **A structured approach to research**: Effective market research involves a structured approach to analyzing trends, market sizing, competitive landscapes, and customer needs. Enhance AI-generated insights by incorporating relevant whitepapers, reports, and customer feedback to provide additional context and uncover unique opportunities.

- **Understand the competition**: Defining the right market category is crucial for differentiation and crafting resonant messaging. Analyze competitor positioning, identify gaps and opportunities, and test potential category labels with customers. Pay attention to how customers describe their problems to their bosses when seeking a budget, as this language can inform powerful category definitions.

- **See the big picture**: Competitive intelligence requires a multidimensional approach, examining competitors' products, positioning, and strategies. Analyze competitor strengths and weaknesses, identify opportunities for differentiation, and simulate competitor perspectives in order to anticipate their moves. Prioritize insights based on market impact, competitive threat, and strategic alignment to focus efforts on the most impactful areas.

Chapter 4 References

[1] Macedo, Matt. *Getting Started with Markdown*. The Markdown Guide. Accessed June 17, 2024. https://www.markdownguide.org/getting-started/.

CHAPTER 5

Audience Understanding

Understanding User Needs

As a PMM, I've found that getting inside your buyers' heads makes all the difference. Let's look at how to really grasp what makes an audience tick.

Start by studying buyer personas closely. You need to understand what keeps them up at night. What are their daily challenges? For example, if working with IT managers, you may find that they're wrestling with cloud migration headaches—not just technical issues but also budget constraints and skill gaps within their team.

A practical approach I use is:
- Pulling insights from customer calls (tools like Gong are great for this).
- Using customer review information from websites like G2Crowd, TrustRadius, Gartner Peer Insights, and any internal feedback surveys.
- Analyze support tickets, feature requests, and Reddit threads.
- Talk directly to sales, solution engineers (SE), and customer success managers (CSM) teams about common customer

pain points. Make sure to record these so you can upload the transcripts into prompts.

Define variables:

- {industry=Insert your target industry, e.g., enterprise SaaS for cybersecurity.}
- {solution=Insert your product/solution, e.g., threat detection platform.}
- {audience=Insert your target audience, e.g., IT security managers, CISOs, or compliance officers.}

Define the objective: Understand user needs

ROLE

You are a user research expert specializing in {industry} and {solution}.

CONTEXT

I need to deeply understand the needs, pain points, and priorities of {audience} in the {industry}. The goal is to gather insights that inform product positioning, feature development, and messaging strategies.

TASK

Analyze the user needs for {audience} by answering the following:

- What are the most common pain points they face related to {solution}?
- What are their primary goals and motivations?
- What challenges or barriers prevent them from achieving these goals?

FORMAT

Provide a structured response with:

- Pain points: List the top 3–5 challenges.
- Goals and motivations: Summarize what they aim to achieve and why.
- Barriers: Highlight the key obstacles preventing success.

TONE

Professional, user-focused, and actionable.

Analyze existing feedback for patterns

ROLE
You are a data analyst helping uncover user needs by analyzing customer feedback.

CONTEXT
I have collected user insights from customer interviews, surveys, reviews, and support tickets. I need to identify recurring patterns, pain points, and emerging needs to refine my understanding of {audience}.

TASK
Analyze the following customer feedback: [Insert data: user quotes, survey responses, reviews, support ticket summaries].
Summarize the most common pain points mentioned.
Identify recurring needs or opportunities that appear across multiple users.
Highlight any unexpected themes or insights.

FORMAT
Provide the response in bullet points:
• Top pain points: Recurring challenges identified.
• User needs: Common themes and expectations.
• Emerging insights: Unexpected patterns or opportunities for innovation.

TONE
Analytical, clear, and insight-driven.

Simulate user interviews

ROLE
You are a customer representing the {audience} in {industry}.

CONTEXT
I am conducting a simulated interview to uncover the day-to-day challenges, priorities, and needs of {audience}. Focus on real-world frustrations and aspirations.

TASK
Answer the following questions as if you are a typical user:
• What keeps you up at night? Describe your biggest

challenges related to {solution}.

- What are your main priorities or goals? What outcomes do you hope to achieve?
- What features or solutions would make your life easier?
- What frustrations do you face with current solutions or tools in the market?

FORMAT

Provide answers in short, user-focused paragraphs, as if responding in an interview.

TONE

Realistic, conversational, and user-centric.

Generate pain point-driven user stories

ROLE

You are a product manager translating user needs into actionable user stories for development.

CONTEXT

I need to create user stories that reflect the pain points, needs, and goals of {audience} for {solution}. These stories will guide feature prioritization and roadmap planning.

TASK

For each identified pain point, create a user story in the following format:

- Pain point: Describe the user's challenge.
- User story: "As a [role], I need [capability] so that I can [achieve a specific goal]." Example:
- Pain point: IT managers spend hours manually resolving system alerts.
- User story: "As an IT manager, I need automated alert triaging so that I can prioritize critical issues and reduce response time."

FORMAT

Provide 3–5 user stories, each mapped to a specific pain point.

TONE

Clear, actionable, and structured.

Identify unmet needs and opportunities

ROLE

You are an industry strategist identifying gaps and opportunities in the market for {solution}.

CONTEXT

Based on user pain points and needs, I want to uncover unmet opportunities where current solutions fall short. This will help position {solution} as a differentiated offering in the market.

TASK

Identify gaps in current solutions and suggest opportunities for innovation:

- What unmet needs are most significant for {audience}?
- How can {solution} address these needs better than competitors?
- Suggest 2–3 feature ideas or enhancements to fill these gaps.

FORMAT

Provide the response in bullet points:

- Unmet needs: Key gaps in current tools or processes.
- Opportunities: Specific ways {solution} can address these gaps.
- Feature ideas: Brief descriptions of proposed features or enhancements.

TONE

Strategic, opportunity-focused, and actionable.

Validate and prioritize needs

ROLE

You are a customer insights analyst validating user needs to prioritize product and messaging strategies.

CONTEXT

I need to validate the identified user needs and prioritize them based on urgency, impact, and alignment with {solution}'s roadmap.

TASK

Evaluate the following list of user needs: [Insert user needs list here].

For each need:

- Rank its importance (High, Medium, Low).
- Assess its impact on user satisfaction or product adoption.
- Suggest whether it should be a short-term or long-term priority.

FORMAT

Provide the response in a table format:

User Need	Importance	Impact	Priority
Example Need 1	High	Critical	Short-Term
Example Need 2	Medium	Moderate	Long-Term

TONE

Structured, analytical, and results-oriented.

Uncover Audience Frustrations (Step 1)

Next, we will move into audience frustrations and identifying unmet needs. Since there is some overlap between them, this will be a two-part prompt:

- **Step 1: Uncover audience frustrations**
  - Focus on reactive pain points and day-to-day struggles.
  - Use existing data like support tickets, reviews, or interview transcripts.
  - **Output**: A list of frustrations ranked by frequency and urgency.
- **Step 2: Identify unmet needs**
  - Analyze frustrations to explore gaps and aspirations beyond immediate problems.
  - Focus on future opportunities for innovation or differentiation.
  - **Output**: A list of unmet needs and innovation opportunities tied to customer goals.

One of my favorite PMM tactics is finding gaps in the market that competitors haven't addressed. Think of it like a puzzle—you're looking for pieces that don't quite fit anywhere else.

Here's my real-world process:

- **Scan customer forums and communities**: I read through Reddit, Stack Overflow, and industry-specific forums. When you see the same complaints popping up repeatedly, you've found a signal worth investigating.
- **Review competitor product reviews**: The three-star reviews are often the most telling. They show what's working AND what's missing.
- **Analyze support tickets**: Look for patterns in feature requests or workarounds that customers create. These highlight where current solutions fall short.

A helpful ChatGPT prompt I use is: "Analyze these common customer complaints and suggest product features that could address the underlying needs: [paste in actual feedback]"

Pro tip:

When you spot an unmet need, validate it across different customer segments. What's a deal-breaker for enterprise clients might be irrelevant for SMBs.

Input variables:

- {industry=Insert your target industry, e.g., enterprise data analytics.}
- {solution=Insert your product/solution, e.g., cloud-based data management platform.}
- {audience=Insert your target audience, e.g., data engineers, IT managers, or compliance officers.}

Define the objective

You are a customer insights analyst specializing in uncovering user pain points and frustrations within {industry}.

</role>

I need to identify the top frustrations that {audience} faces when using {solution} or similar tools. This will help align messaging, prioritize feature fixes, and ensure we address the most critical pain points for our customers.

</context>

Analyze the following data sources to uncover recurring audience frustrations:

- **Customer feedback**: Reviews, surveys, or interview transcripts.
- **Support tickets**: Summaries of user-reported problems and complaints.
- **Online communities**: Forums, Reddit threads, or social media comments related to {solution} or competitors.

Summarize the findings by:

- Listing the top 3–5 frustrations with clear descriptions.
- Highlighting recurring themes or patterns across different sources.
- Identifying any critical pain points that are urgent or widespread.

</task>

<format>

Provide the response as follows:

1. Top frustrations:
   o Frustration 1: [Short description of the problem].
   o Frustration 2: [Short description of the problem].
   o Frustration 3: [Short description of the problem].
2. Recurring patterns:
   o [Summarize common themes observed across feedback].
3. Critical pain points:
   o [Highlight which frustrations are the most urgent or widespread].

</format>

<tone>
Analytical, user-focused, and concise.
</tone>

Analyze existing customer feedback

You are a research analyst summarizing patterns in customer feedback to identify audience frustrations.
</role>

I have raw customer feedback collected from reviews, surveys, and transcripts. I need you to analyze this data and extract recurring pain points or frustrations expressed by {audience}.
</context>

Analyze the following customer feedback: [Insert customer reviews, survey responses, or transcripts here.]
• Identify the most common frustrations or complaints.
• Group similar frustrations into themes or categories.
• Highlight any unexpected insights or pain points that stand out.
</task>

<format>
Provide the response in bullet points under these headings:
• Recurring frustrations:
    o [List the top frustrations with a brief description].
• Thematic categories:
    o [Group the related frustrations under clear themes, e.g., "Onboarding Challenges," "Integration Issues," or "Performance Limitations"].
• Unexpected insights:
    o [Highlight any surprising or less obvious frustrations].
</format>

<tone>
Structured, clear, and insight-driven.
</tone>

Simulate user feedback (if data is limited)

You are a typical user of {solution} in {industry}. You represent the {audience} who faces daily challenges with data tools.
</role>
I am trying to uncover audience frustrations, but I have limited real-world feedback. I need you to simulate realistic complaints and pain points that {audience} might experience when using {solution}.
</context>
Answer the following questions as if you are a user experiencing frustrations with {solution}:
- **What challenges do you face during onboarding or setup?**
- **What are the most common issues you encounter while using {solution}?**
- **What tasks or workflows feel inefficient, slow, or frustrating?**
- **What do you find frustrating about the support, documentation, or user interface?**
</task>
<format>
Provide answers in short, user-focused paragraphs as realistic responses to the above questions.
</format>
<tone>
Conversational, realistic, and frustration-driven.
</tone>

Prioritize frustrations based on impact

You are a strategist helping prioritize user frustrations to guide product improvements and messaging alignment.

</role>
I have identified a list of user frustrations for {audience}. I need you to prioritize them based on their frequency, severity, and impact on customer satisfaction.
</context>
Prioritize the following frustrations based on:
1. Frequency: How often does this issue occur?
2. Severity: How disruptive is it to the user's workflow?
3. Impact: How does this frustration affect overall satisfaction with {solution}?
Here's the list of frustrations:
[Insert identified frustrations here.]
</task>
<format>
Provide the response in a table format:

Frustration	Frequency	Severity	Impact	Priority
Example: Integration difficulties	High	Critical	Affects workflows	High
Example: UI navigation issues	Medium	Moderate	Slows efficiency	Medium

</format>
<tone>
Strategic, analytical, and prioritized.
</tone>

Summarize and align with next the steps

You are a product marketer summarizing audience frustrations for the product and messaging teams.
</role>

I need to summarize the most critical user frustrations and align them with actionable next steps for product and marketing teams.

</context>

Summarize the findings as follows:

- Top 3 recurring audience frustrations.
- Key themes or categories these frustrations fall into.
- Recommended next steps to address these frustrations in product improvements, messaging, or support strategies.

</task>

<format>

Provide the response in a structured format:

1. Top frustrations:
    o Frustration 1: [Description].
    o Frustration 2: [Description].
    o Frustration 3: [Description].

2. Themes:
    o [List overarching themes].

3. Next steps:
    o Product improvement: [Recommendation].
    o Messaging alignment: [Recommendation].
    o Support strategy: [Recommendation].

</format>

<tone>

Executive-level, actionable, and solution-focused.

</tone>

Identify Unmet Needs (Step 2)

Building upon the frustrations identified in Step 1, we will now uncover opportunities for innovation and differentiation.

Define the objective

You are a product strategist specializing in uncovering unmet customer needs within {industry}.

</role>

I need to identify **unmet needs**—the gaps or opportunities where existing solutions fail to fully address the challenges of {audience}. These insights will inform innovation, product strategy, and positioning for {solution}.

</context>

Based on the audience's frustrations identified earlier, determine:

- What **gaps** exist in current solutions that leave user needs unfulfilled?
- What **aspirational goals** or desired outcomes do users have that are not being met?
- What opportunities exist for {solution} to address these gaps and stand out from competitors?

</task>

<format>

Provide the response as follows:

1. Unmet needs:
   o Need 1: [Brief description of the gap].
   o Need 2: [Brief description of the gap].
2. Aspirational goals:
   o Goal 1: [What users want to achieve beyond their current limitations].
   o Goal 2: [A forward-looking need that competitors fail to address].
3. Opportunities for {solution}:
   o Opportunity 1: [How {solution} can address this unmet need].
   o Opportunity 2: [A specific feature, capability, or innovation to close the gap].

</format>
<tone>
Strategic, opportunity-focused, and user-centric.
</tone>

Analyze gaps in current solutions

<role/>
You are an industry analyst comparing {solution} to existing tools in {industry} to identify unmet customer needs.
</role>
<context/>
I want to understand where current solutions fail to meet the needs of {audience}. Focus on identifying specific areas where competitors underperform or fail to deliver desired outcomes.
</context>
<task/>
Analyze the following aspects of existing solutions:
- **Feature gaps**: What critical features or capabilities are missing in current tools?
- **Experience gaps**: What processes or workflows remain inefficient, complex, or manual?
- **Outcome gaps**: What desired outcomes do users fail to achieve despite using existing solutions?
</task>
<format>
Provide the response in bullet points under these categories:
- Feature gaps:
    o Gap 1: [Short description].
    o Gap 2: [Short description].
- Experience gaps:
    o Gap 1: [What part of the workflow is frustrating or inefficient].
    o Gap 2: [Another experience-related gap].
- Outcome gaps:
    o Gap 1: [The result users want but don't achieve].

o Gap 2: [Another unmet outcome].
</format>
<tone>
Analytical, clear, and actionable.
</tone>

Simulate aspirational needs through user interviews

You are a user representing the {audience} in {industry}.
</role>
I need to understand the aspirational goals and unmet needs of {audience} beyond their current frustrations. Focus on forward-looking outcomes they wish to achieve with {solution}.
</context>
Answer the following questions as if you are a user describing aspirational needs:
- **What would an ideal solution enable you to achieve that current tools cannot?**
- **What processes or tasks do you wish were faster, automated, or simplified?**
- **What outcomes or successes would make your job significantly easier or more valuable?**
- **What innovations would make you excited about using {solution}?**
</task>
<format>
Provide responses as short, user-focused paragraphs that reflect aspirational needs and desired outcomes.
</format>
<tone>
Conversational, forward-looking, and aspirational.
</tone>

Prioritize unmet needs based on impact

You are a strategist prioritizing unmet needs to guide innovation and product development for {solution}.

</role>

I have identified a list of unmet needs for {audience} in {industry}. I need to prioritize them based on their potential business impact, urgency, and feasibility of addressing them.

</context>

Prioritize the following unmet needs based on:

- **Impact**: How significant is this need for the user?
- **Urgency**: How pressing is it for users to solve this issue?
- **Feasibility**: How achievable is it for {solution} to address this need?

Here's the list of unmet needs:

[Insert list of unmet needs here.]

</task>

<format>

Provide the response in a table format:

Unmet Need	Impact	Urgency	Feasibility	Priority
Example: Faster automation	High	High	Medium	High Priority
Example: Enhanced insights	Medium	Low	High	Medium Priority

</format>

<tone>

Strategic, structured, and actionable.

</tone>

Summarize unmet needs and opportunities

You are a senior product strategist presenting insights on unmet needs and opportunities for {solution} to executive stakeholders.

</role>

Summarize the identified unmet needs and highlight the most significant opportunities for innovation and differentiation in the {industry}.

</context>

Provide a concise summary as follows:

- **Top unmet needs**: List the most critical needs that current tools fail to address.
- **Opportunities for {solution}**: Highlight 2–3 actionable opportunities to close these gaps.
- **Strategic recommendations**: Suggest next steps to address these opportunities, including feature ideas, messaging strategies, or competitive positioning.

</task>

<format>

Provide the response in a structured format:

1. Top unmet needs:
   o Need 1: [Short description].
   o Need 2: [Short description].
2. Opportunities:
   o Opportunity 1: [How {solution} can address this need].
   o Opportunity 2: [Innovation or feature recommendation].
3. Strategic recommendations:
   o [Actionable suggestion 1].
   o [Actionable suggestion 2].

</format>

Executive-level, clear, and results-driven.

</tone>

Pro tip:

Always validate personas by asking customer-facing teams like account executives, sales engineers, and customer success managers to review them. They're often closest to the economic and technical buyers and can spot if something important is being missed.

Voice of Customer

Understanding the voice of the customer is where we really get into the heads of our buyers, recognize their needs, and fine-tune our messaging.

Now, we all know there are tons of ways to gather customer feedback—surveys, interviews, focus groups, social media listening. But what happens when you're drowning in data? How do you make sense of it all?

Instead of manually sifting through pages of survey responses or hours of interview transcripts, use AI to do the heavy lifting.

Here's how: Feed the AI the raw customer feedback data, anything from survey results to customer support tickets. Then, use a prompt like this:

Define input variables:

- {industry=Insert your target industry, e.g., cloud-based cybersecurity solutions.}
- {solution=Insert your product/solution, e.g., automated threat detection platform.}
- {audience=Insert your target audience, e.g., IT security managers, SOC analysts, or CISOs.}

Define the objective

You are a customer insights analyst specializing in synthesizing feedback to uncover actionable themes and insights for {solution} in the {industry}.
</role>
I have collected customer feedback from various sources, including surveys, interviews, reviews, and support tickets. I need you to analyze and synthesize this data to identify:

- Key pain points expressed by {audience}.
- Top priorities, needs, and goals.
- Recurring themes across feedback sources.
- Opportunities for product improvement and messaging refinement.

</context>
Analyze the following customer feedback data: [Insert data: customer interviews, survey responses, reviews, or support ticket summaries.]

- Identify recurring pain points or frustrations.
- Highlight top priorities, needs, and goals expressed by users.
- Group feedback into thematic categories.
- Provide actionable insights for product improvements or refined messaging.

</task>
<format>
Provide the response in a structured format:

1. Recurring pain points:
    o Pain Point 1: [Description].
    o Pain Point 2: [Description].
2. Top priorities and goals:
    o Priority 1: [Description].
    o Priority 2: [Description].
3. Thematic categories:
    o Theme 1: [Cluster of related feedback].

o Theme 2: [Cluster of related feedback].
4. Actionable insights:
   o Insight 1: [Recommendation for product improvement].
   o Insight 2: [Suggestion for messaging refinement].
</format>
<tone>
Analytical, user-focused, and actionable.
</tone>

Analyze feedback across channels

You are a VoC expert analyzing multichannel feedback to uncover patterns and trends.
</role>
I have collected customer feedback from multiple sources, including:
- Online reviews (e.g., G2, TrustRadius).
- Survey responses.
- Customer interviews.
- Support tickets and chat logs.

I need you to synthesize insights across these sources and identify common themes, discrepancies, and opportunities.
</context>
For each feedback channel:
1. Summarize the most frequent pain points and needs mentioned.
2. Identify any differences or discrepancies in feedback between channels.
3. Combine insights into unified themes that apply across all channels.
</task>

Provide the response as follows:
- Channel summaries:
  - Surveys: [Top pain points, needs, and patterns].
  - Reviews: [Key frustrations, priorities, and recurring themes].
  - Support Tickets: [Issues most frequently reported].
  - Interviews: [Deeper insights into user challenges and goals].
- Cross-channel themes:
  - Theme 1: [Description of theme and supporting examples].
  - Theme 2: [Description of theme and supporting examples].
- Discrepancies:
  - [Highlight differences in feedback between channels and why they might exist].
</format>
<tone>
Structured, comparative, and insight-driven.
</tone>

Simulate customer perspectives for gaps

You are a typical user of {solution} in {industry}, sharing perspectives about your frustrations, goals, and unmet needs.
</role>
To ensure we fully understand customer sentiment, I need to simulate customer perspectives to identify any gaps not surfaced in our current VoC data.
 </context>
Answer the following questions as if you are a user of {solution}:

- **What are your biggest frustrations when using solutions like {solution}?**
- **What outcomes or goals are most important to you when evaluating this type of solution?**
- **What features or improvements would make this solution significantly better for you?**
- **What other tools or workarounds do you use to compensate for gaps in current solutions?**

</task>

<format>

Provide short, user-focused paragraphs for each question, simulating realistic feedback.

</format>

<tone>

Conversational, authentic, and user-centric.

</tone>

Prioritize VoC themes for actionability

You are a strategist prioritizing VoC themes to inform product roadmap and messaging.

</role>

I need to prioritize the synthesized VoC insights based on their urgency, impact, and alignment with {solution}'s business goals.

</context>

Prioritize the identified VoC themes using the following criteria:

- **Frequency**: How often the issue or need is mentioned.
- **Impact**: The severity of the issue or its importance to the customer's workflow.
- **Alignment**: How well addressing this theme aligns with our product strategy. </task>

Provide the response in a table format:

Theme	Frequency	Impact	Alignment	Priority
Example: Integration Issues	High	Critical	Strong	High Priority
Example: Onboarding Support	Medium	Moderate	Medium	Medium Priority

</format>
<tone>
Analytical, structured, and prioritization-focused.
</tone>

Summarize VoC insights and recommendations

You are a senior product strategist presenting VoC insights to inform product and marketing decisions for {solution}.
</role>
I need a concise summary of the synthesized Voice of Customer insights, highlighting key findings and actionable recommendations for product improvements and messaging strategies.
</context>
Summarize the VoC synthesis as follows:
- **Top themes**: The most critical themes or pain points identified.
- **Customer priorities**: Key goals and needs expressed by the audience.
- **Opportunities**: Areas where {solution} can address gaps or differentiate.
- **Recommendations**: Actionable next steps for product, messaging, and support strategies.

</task>
<format>
Provide the response in the following structure:
1. Top themes:
   o Theme 1: [Description].
   o Theme 2: [Description].
2. Customer priorities:
   o Priority 1: [Brief explanation].
   o Priority 2: [Brief explanation].
3. Opportunities:
   o Opportunity 1: [Innovation or solution area].
   o Opportunity 2: [Differentiation suggestion].
4. Recommendations:
   o Product Improvement: [Actionable suggestion].
   o Messaging Strategy: [Actionable suggestion].
   o Support Enhancements: [Actionable suggestion].
</format>
<tone>
Executive-level, strategic, and actionable.
</tone>

See how that works? You're giving the AI clear instructions on what to do with the data.

And here's where it gets really interesting. The AI can then aggregate that feedback into actionable categories. Maybe you'll discover that security is a top concern, or ease of integration is a major priority. Such insights can then be used to refine messaging and positioning.

The goal is to make sure your message resonates with the target audience. AI can help do this by quickly synthesizing large volumes of customer feedback.

Win-Loss Analysis

Okay, let's get real for a minute. We all know that not every deal goes our way. But losses can be incredibly valuable if we use them to learn and improve. That's where win-loss analysis comes in.

Traditionally, this involves a lot of manual work. Interviewing sales reps, reviewing notes, and trying to piece together why we won or lost a deal. But AI can help streamline this process and uncover hidden insights.

Imagine feeding the AI your CRM data and transcripts from sales calls (tools like Gong are great for this), then using a prompt like this:

Define your inputs:

- {industry=Insert your target industry, e.g., enterprise cybersecurity solutions.}
- {solution=Insert your product/solution, e.g., threat detection platform.}
- {audience=Insert your target audience, e.g., CISOs, IT security managers, procurement teams.}
- {data=Insert required data sources, e.g., CRM deal records, sales interview notes, customer feedback, post-mortem reports.}

Define the objective

You are a win-loss analysis expert helping a product marketing team uncover why deals are won or lost for {solution} in the {industry}.

</role>

I need to analyze win-loss data to identify patterns, reasons for success or failure, and opportunities for improving our sales process, product messaging, and positioning. The goal is to refine strategy, better align with customer needs, and address competitive challenges.

</context>

Using the following win-loss data: [Insert data sources: CRM deal outcomes, sales call notes, customer feedback, and reasons tagged in closed/lost records.]

- Identify the top reasons deals are won (e.g., product features, pricing, support).
- Identify the top reasons deals are lost (e.g., competitor strengths, objections, pricing concerns).
- Summarize recurring themes or patterns in the data.
- Provide actionable insights and recommendations for sales enablement, messaging improvements, and product development.

</task>
<format>
Provide the response as follows:
1. Top reasons deals were won:
   o Reason 1: [Description].
   o Reason 2: [Description].
2. Top reasons deals were lost:
   o Reason 1: [Description].
   o Reason 2: [Description].
3. Recurring themes:
   o Theme 1: [Patterns observed].
   o Theme 2: [Patterns observed].
4. Actionable recommendations:
   o Recommendation 1: [Sales enablement or messaging improvement].
   o Recommendation 2: [Product feature or competitive adjustment].
</format>
<tone>
Professional, analytical, and results-driven.
</tone>

Analyze sales call notes and customer feedback

You are a customer insights analyst summarizing patterns from sales call notes and customer feedback related to won and lost deals.
</role>

I need you to analyze qualitative data, including sales call transcripts and post-mortem feedback, to extract recurring reasons for deal outcomes. Focus on identifying why customers chose or rejected {solution}.
</context>
Analyze the following notes and feedback: [Insert data: sales call transcripts, customer quotes, survey feedback.]
- For won deals: Identify common themes or statements about what influenced the decision to purchase.
- For lost deals: Identify recurring objections, concerns, or mentions of competitors that led to the loss.
- Highlight any patterns related to product features, pricing, messaging, or sales process.
</task>
<format>
Provide the response in bullet points:
- Reasons for won deals:
    o [Key factor 1].
    o [Key factor 2].
- Reasons for lost deals:
    o [Key factor 1].
    o [Key factor 2].
- Recurring patterns:
    o [Theme 1: e.g., "Pricing concerns were most common among SMBs."]
    o [Theme 2: e.g., "Competitor X's ease of integration was frequently mentioned."]
</format>
<tone>
Insightful, detailed, and pattern-focused.
</tone>

Simulate competitor and customer perspectives

You are simulating a customer or competitor to help surface additional insights into win-loss factors for {solution}.
</role>
To enhance the win-loss analysis, I need to simulate both customer perspectives (why they chose or rejected us) and competitor perspectives (how they position themselves to win deals).
</context>
Answer the following questions as if you are:
- **A customer who chose {solution}**:
  o What key factors led you to select {solution}?
  o What specific features or outcomes were most valuable?
- A customer who chose a competitor:
  o Why did you prefer the competitor's solution?
  o What concerns or gaps caused you to reject {solution}?
- A competitor in {industry}:
  o How do you position your product to win deals against {solution}?
  o What weaknesses or gaps in {solution} do you highlight to prospects?
</task>
<format>
Provide responses as short, role-played paragraphs for each perspective:
- Customer (won): [Realistic response].
- Customer (lost): [Realistic response].
- Competitor: [Realistic competitive positioning response].
</format>
<tone>
Realistic, role-specific, and strategic.
</tone>

Prioritize win-loss insights

You are a product strategist prioritizing win-loss insights based on their impact and frequency.

</role>

I need to prioritize the reasons for won and lost deals to inform sales strategy, product improvements, and competitive positioning.

</context>

Prioritize the win-loss insights based on the following criteria:

- **Frequency**: How often is this factor mentioned across deals?
- **Impact**: How significant is this factor in influencing the final decision?
- **Strategic importance**: How well does addressing this factor align with {solution}'s business goals?

</task>

<format>

Provide the response in a table format:

Reason/ Insight	Frequency	Impact	Strategic Importance	Priority
Example: Pricing too high	High	Critical	Strong	High Priority
Example: Competitor X feature	Medium	Moderate	Medium	Medium Priority

</format>

<tone>

Structured, strategic, and prioritized.

</tone>

Summarize insights and recommendations

You are a senior strategist summarizing win-loss findings to help refine sales, messaging, and product strategy for {solution}.
</role>
Summarize the findings from win-loss analysis and provide actionable recommendations for improving performance in future deals.
</context>
Provide a summary of insights and recommendations as follows:

- **Top reasons for wins**: What factors consistently drive success?
- **Top reasons for losses**: What recurring objections or challenges result in lost deals?
- **Key themes**: Overarching patterns or trends from the analysis.
- **Recommendations**: Specific actions to address weaknesses, improve positioning, or enable sales success.

</task>
<format>
Provide the response in a structured format:

1. Top reasons for wins:
   o Reason 1: [Description].
   o Reason 2: [Description].
2. Top reasons for losses:
   o Reason 1: [Description].
   o Reason 2: [Description].
3. Key themes:
   o Theme 1: [Description].
   o Theme 2: [Description].
4. Recommendations:
   o Sales Enablement: [Actionable suggestion].
   o Messaging Improvement: [Actionable suggestion].
   o Product Adjustment: [Actionable suggestion].

</format>

<tone>

Executive-level, concise, and action-oriented.

</tone>

Why this workflow works

- **Holistic analysis**: Covers qualitative and quantitative data from multiple sources.
- **Customer and competitor insights**: Simulates perspectives to surface gaps and opportunities.
- **Actionable outputs**: Delivers prioritized recommendations for sales enablement, product development, and competitive positioning.
- **Strategic alignment**: Ensures insights align with business goals and customer needs.

By following this workflow, PMMs can use ChatGPT to synthesize win-loss data, identify actionable insights, and build strategies to improve deal performance and product positioning.

Practice Exercises

- **Conduct a user needs analysis**: Define the industry, solution, and target audience, then step through the prompts to identify common pain points, goals, motivations, and barriers to success. Analyze existing customer feedback to uncover patterns, simulate user interviews to gain deeper insights, and generate pain point-driven user stories. Identify unmet needs and opportunities for innovation, then validate and prioritize the needs based on importance and impact. Summarize findings in a presentation for your product and marketing teams.
- **Practice uncovering audience frustrations**: Identify the unmet needs of your product or a hypothetical one. Follow the two-step process outlined in the chapter. In Step 1, analyze customer feedback from various sources (e.g., forums, reviews, support tickets) to identify recurring complaints and pain points. Prioritize the frustrations based

on frequency, severity, and impact on user satisfaction. In Step 2, explore gaps in current solutions and aspirational goals that are not being met. Simulate user interviews to uncover forward-looking needs, then prioritize the unmet needs based on impact, urgency, and feasibility. Summarize findings and provide strategic recommendations for product improvements and positioning.

- **Conduct a win-loss analysis**: For a product or a hypothetical one, define the target industry, solution, audience, and relevant data sources (e.g., CRM records, sales call notes, customer feedback). Analyze the data to identify the top reasons deals were won or lost, recurring themes, and actionable insights for improvement. Simulate competitor and customer perspectives to surface additional factors influencing deal outcomes. Prioritize the win-loss insights based on frequency, impact, and strategic importance. Summarize findings and provide recommendations for sales enablement, messaging refinement, and product adjustments to improve future deal performance.

Summary

- **Study the end users**: Effective user needs analysis involves studying buyer personas, analyzing customer feedback, simulating user interviews, and generating pain point-driven user stories. Identify unmet needs and opportunities for innovation by comparing current solutions to aspirational goals, then validate and prioritize needs based on importance and impact.

- **Understand what users want**: Uncovering audience frustrations requires analyzing customer feedback from various sources, such as forums, reviews, and support tickets, and identifying recurring pain points. Prioritize frustrations based on frequency, severity, and impact on user satisfaction. By understanding these frustrations you can explore gaps in current solutions and uncover unmet needs and opportunities for differentiation.

- **Get a handle on winning and losing**: Conducting win-loss analysis involves examining CRM records, sales call notes, and customer feedback to identify top reasons for won and lost deals. Simulate competitor and customer perspectives to surface additional insights, then prioritize findings based on frequency, impact, and strategic importance. Provide actionable recommendations for sales enablement, messaging refinement, and product improvements to enhance future deal performance.

CHAPTER 6

Positioning and Messaging

Positioning Framework Creation

Positioning is how you frame a product in the context of the market. In Crossing the Chasm, Moore explains:

- Positioning, first and foremost, is a noun, not a verb.
- Positioning is the single largest influence on the buying decision.
- Positioning exists in people's heads, not in your words.
- People are highly resistant to changes in positioning.

The analyst firm Gartner defines positioning as: "An effective positioning statement is the crystallization of your value proposition, differentiated for a specific target audience."[10]

You can download a messaging framework template adapted from Moore's original work at TinyTechGuides.com/Templates.

Define the inputs:

- {industry=Insert your target industry, e.g., enterprise cloud security.}
- {solution=Insert your product/solution, e.g., cloud-based threat detection platform.}
- {target audience=Insert primary audience, e.g., CISOs, IT security leaders, DevOps engineers.}

- {competition=Insert key competitors, e.g., Competitor A, Competitor B.}

Define the objective

You are a product marketing expert specializing in positioning strategies for {industry}.
</role>
I need to develop a positioning framework for {solution} using Geoffrey Moore's Positioning Model. This will help clarify our target audience, differentiators, and value proposition relative to competitors.
</context>
Based on Moore's framework, create a positioning statement that answers the following:

- **Target customer**: Who is the ideal customer for {solution}?
- **Problem**: What pressing problem or challenge does the target customer face? 3. **Product Category**: What category does {solution} belong to?
- **Unique value**: What is the key differentiator or benefit {solution} offers?
- **Competition**: Who are the key competitors, and how is {solution} better?

Combine these into the following structure:
"For [target customer], who face [key problem], [solution] is a [product category] that provides [unique value]. Unlike [competition], our product [key differentiator]."
</task>
<format>
Provide the completed positioning statement and include a breakdown of each component for clarity:
1. Target Customer: [Brief description].
2. Problem: [The specific challenge being addressed].

3. Product Category: [The category {solution} fits into].

4. Unique Value: [The primary differentiator or key benefit].

5. Competition: [Key competitors and how {solution} outperforms them].

6. Final Positioning Statement:

"For [target customer], who face [key problem], [solution] is a [product category] that provides [unique value]. Unlike [competition], our product [key differentiator]."

</format>

<tone>

Clear, structured, and market-focused.

</tone>

Identify the target customer and problem

You are a customer insights analyst defining the target audience and their core challenges for {solution}.

</role>

To build the positioning framework, I need to clearly articulate the target customer and the problem they face. Focus on identifying the audience that will benefit most from {solution} and the specific challenge they encounter.

</context>

Answer the following questions:

- Who is the **primary target audience** for {solution}? Include role/title, company size, and industry specifics.
- What is the **key problem** they face that {solution} can solve? Describe it in simple terms.
- How does this problem impact their work or business outcomes?

</task>

<format>

Provide the response in bullet points:

- Target Customer:

- o [role/title, e.g., "CISO at large financial enterprises"].
- o [Company size, e.g., "Organizations with 2,000+ employees"].
- o [Industry, e.g., "Financial services with strict compliance needs"].
- Problem:
  - o [Description of the key challenge, e.g., "Managing real-time security threats across multi-cloud environments"].
- Impact:
  - o [The consequence of the problem, e.g., "Increased risk of data breaches and regulatory penalties"].

</format>
<tone>
User-focused and problem-centric.
</tone>

Define the product category and unique value

You are a positioning strategist clarifying the product category and unique differentiators for {solution}.
</role>
I need to define the product category for {solution} and articulate its unique value to ensure clarity and differentiation in the market. Focus on positioning {solution} in a way that aligns with customer expectations while standing out from competitors.
</context>

- Identify the **product category**: What broader category does {solution} fit into (e.g., cloud security platforms, data governance tools)?
- Define the **unique value**: What is the primary benefit {solution} delivers that differentiates it from others in this category?

- Describe the **key differentiator**: What specific feature, capability, or approach sets {solution} apart?

</task>

<format>

Provide the response as follows:

- Product category: [Short description of the category].
- Unique value: [The core benefit or outcome delivered, e.g., "Real-time threat detection with automated response workflows"].
- Key differentiator: [What makes {solution} unique, e.g., "AI-driven insights that reduce false positives by 50%"].

</format>

<tone>

Clear, differentiator-focused, and concise.

</tone>

Analyze competitors to refine differentiation

You are a competitive intelligence analyst comparing {solution} to {competitors} in {industry}.

</role>

To strengthen the positioning framework, I need to understand the competitive landscape and clarify how {solution} outperforms {competitors}. Focus on their strengths and weaknesses relative to {solution}.

</context>

Analyze the positioning of the following competitors: [Insert list of competitors]. For each competitor:

- Summarize their positioning and key value proposition.
- Identify their main strengths.
- Identify their gaps or weaknesses compared to {solution}.
- Explain how {solution} differentiates itself from this competitor.

</task>

Provide the response in bullet points for each competitor:
• Competitor Name:
    o Positioning: [Summary of their positioning].
    o Strengths: [Key advantages they emphasize].
    o Weaknesses: [Gaps in their product or offering].
    o Our Differentiation: [How {solution} stands out].
</format>
<tone>
Analytical, competitive, and focused on differentiation.
</tone>

Finalize the positioning statement

You are a product marketing strategist finalizing the positioning statement for {solution}.
</role>
Based on the identified target customer, problem, product category, unique value, and competitive differentiation, I need a polished, final positioning statement.
</context>
Craft the final positioning statement using this structure:
"For [target customer], who face [key problem], [solution] is a [product category] that provides [unique value]. Unlike [competition], our product [key differentiator]."
</task>
<format>
Provide the completed positioning statement:
• **Positioning statement**: [Insert final positioning statement].
Provide a short explanation for each component to ensure clarity.

<tone>
Clear, confident, and market-ready.
</tone>

Why this workflow works

- **Aligns with Moore's framework**: Breaks down positioning into clear, logical components (target customer, problem, product category, unique value, and competition).
- **Step-by-step clarity**: Guides PMMs through identifying inputs, analyzing competitors, and refining the final statement.
- **Actionable outputs**: Ensures deliverables are structured and immediately usable.
- **Competitive differentiation**: Integrates competitor analysis to highlight unique strengths.

By following this workflow, PMMs can leverage ChatGPT to create a clear, compelling positioning framework that resonates with their target audience while standing out in a competitive market.

Messaging Framework Creation

Creating effective messaging frameworks doesn't have to be complicated. In my experience, they work best when answering three basic questions: who needs our product, what problems does it solve, and why should customers choose us over alternatives.

If you visit TinyTechGuides.com/templates, you can download the messaging template and simply ask GPT this question:

Prompt:

Ok, using my messaging template, can you help fill this out?

Elevator pitch and points of distinction elevator pitch:

<e.g. Your Elevator Pitch or Tagline>

Points of distinction:

< 3–5 value-based headlines, with three supporting key points; write messages for the different audiences you serve.

These should be MECE (mutually exclusive and collectively exhaustive)>

Headline #1: TBD 1 TBD 2 TBD 3

Headline #2: TBD 1 TBD 2 TBD 3

Headline #3: TBD 1 TBD 2 TBD 3

Copy blocks (to be completed after the points of distinction):

5 Word Description: TBD

25 Word Description: TBD

50 Word Description: TBD

Social Media: TBD 1 TBD 2 TBD 3

Now onward to the precise workflow.

Define the inputs

- {industry=Insert your target industry, e.g., cloud infrastructure for enterprises.}
- {solution=Insert your product/solution, e.g., automated cloud cost optimization platform.}
- {audience=Insert target audience, e.g., IT operations leaders, CFOs, DevOps engineers.}
- {value drivers=Insert key value drivers, e.g., cost savings, performance optimization, and workflow automation.}

Define the objective

You are a product marketing expert creating a messaging framework for {solution} in the {industry}.

</role>

I need to develop a messaging framework that clearly articulates the value of {solution} to {audience}. This framework will support sales, marketing, and customer success efforts by providing consistent, audience-specific messaging that highlights our unique value proposition and key differentiators.

</context>

Create a messaging framework for {solution} that includes the following:

- **Target audience**: Who are we speaking to (roles, industries, pain points)?
- **Core value proposition**: The overarching value delivered by {solution}.
- **Key messages**: The top 3–5 messages that convey the value of {solution} to the audience.
- **Supporting proof points**: Features, benefits, or data that reinforce each key message.

</task>
<format>
Provide the response in a structured format:

1. Target audience:
   - role: [e.g., IT leaders, CFOs].
   - Key Pain Points: [e.g., "Difficulty managing cloud costs," "Lack of visibility into infrastructure performance"].
2. Core value proposition:
   - "[Solution] helps [audience] achieve [outcome] by [key differentiator]."
3. Key messages and proof points:

Key Message	Supporting Proof Points
[Message 1: Key value communicated]	[Feature or benefit 1]
[Message 2: Key value communicated]	[Feature or benefit 2]
[Message 3: Key value communicated]	[Feature or benefit 3]

</format>
<tone>
Clear, concise, and customer-focused.
</tone>

Identify the target audience and pain points

You are a customer research specialist defining the target audience and their key challenges for {solution}.
</role>

To ensure our messaging resonates, I need to clearly identify the primary audience and their most pressing pain points. Focus on the specific roles, industries, and challenges they face that {solution} can address.
</context>

Answer the following:

- Who is the **primary target audience** for {solution}? Include details like role, company size, and industry.
- What are their **top 3 pain points** that {solution} can solve?
- How do these pain points impact their business or workflows?

</task>

<format>

Provide the response in bullet points:

- Target audience:
  - Role: [Primary roles/titles].
  - Company size: [SMB, mid-market, enterprise].
  - Industry: [Industry specifics].
- Top pain points:
  - Pain Point 1: [Brief description of the challenge].
  - Pain Point 2: [Brief description of the challenge].
  - Pain Point 3: [Brief description of the challenge].
- Impact:
  - [How the pain points affect business outcomes or workflows].

</format>

<tone>

Audience-focused, empathetic, and problem-driven.
</tone>

Develop the core value proposition

You are a positioning strategist crafting a compelling core value proposition for {solution}.
</role>
I need a clear and concise core value proposition that communicates how {solution} solves the audience's key pain points and what differentiates it from competitors.
</context>
Write a core value proposition in the following format: "[Solution] helps [audience] achieve [specific outcome or benefit] by [key differentiator or unique capability]."
Include a short explanation of why this value matters to the audience.
</task>
<format>
Provide the core value proposition as follows:
- Core value proposition: "[Solution] helps [audience] achieve [specific outcome] by [key differentiator]."
- Why it matters: [Brief explanation of the relevance to the audience's goals or challenges].
</format>
<tone>
Clear, benefit-driven, and compelling.
</tone>

Create key messages and supporting proof points

You are a messaging expert developing key messages and proof points to communicate the value of {solution}.
</role>

I need 3–5 clear key messages that articulate the value of {solution} to the target audience. Each message must be supported with proof points such as features, benefits, data, or customer outcomes.

</context>

Develop key messages and supporting proof points:
- **Key message**: What is the main value or benefit?
- **Supporting proof points**: What evidence (e.g., product features, data, testimonials) supports this message?

</task>

<format>

Provide the response in a table format:

Key Message	Supporting Proof Points
Message 1: [Value communicated]	Feature/Benefit 1: [Short description]
	Feature/Benefit 2: [Short description]
Message 2: [Value communicated]	Feature/Benefit 1: [Short description]
Message 3: [Value communicated]	Feature/Benefit 1: [Short description]

</format>

<tone>

Crisp, benefit-oriented, and results-focused.

</tone>

Align messaging for different audiences

You are a product marketing strategist tailoring key messages for different audience segments within {industry}.

</role>

The target audience for {solution} includes multiple roles or industries. I need you to align the messaging framework to address the specific priorities and pain points of each audience segment.

</context>

Tailor the key messages for each audience segment:

- List the main audience segments.
- For each segment, adjust the key messages to align with their pain points and priorities.

</task>

<format>

Provide responses as follows:

- Audience Segment 1: [e.g., IT Operations Leaders]
    o Key Message 1: [Revised message for this audience].
    o Key Message 2: [Revised message for this audience].
- Audience Segment 2: [e.g., CFOs]
    o Key Message 1: [Revised message for this audience].
    o Key Message 2: [Revised message for this audience].

</format>

<tone>

Audience-specific, clear, and adaptable.

</tone>

Summarize the messaging framework

You are a senior PMM summarizing the finalized messaging framework for {solution} to support sales, marketing, and customer success teams.

</role>

Summarize the finalized messaging framework in a single, actionable document that can be shared with stakeholders.

</context>

Summarize the messaging framework as follows:

- Target audience: [Who the messaging is for].
- Core value proposition: [The overarching value].
- Key messages and proof points: A structured list of the key messages and supporting evidence.
- Audience-specific messaging: Tailored messages for each key segment.

</task>

<format>

Provide the response as a summary document with clear sections.

1. Target audience: [Description].
2. Core value proposition: "[Solution] helps [audience] achieve [specific outcome] by [key differentiator]."
3. Key messages:

Key Message	Proof Points
Message 1	[Supporting data/feature 1]
Message 2	[Supporting data/feature 2]

4. Audience-specific messaging:
    o Segment 1: [Key messages tailored to Segment 1].
    o Segment 2: [Key messages tailored to Segment 2].

</format>

<tone>

Structured, actionable, and ready-to-implement.

</tone>

Why this workflow works

- **Step-by-step**: Guides PMMs through identifying audiences, value propositions, and key messages.
- **Audience alignment**: Ensures messaging is tailored to specific customer needs.
- **Structured outputs**: Provides actionable deliverables for sales, marketing, and product teams.
- **Focused on value**: Keeps the messaging centered on benefits, proof points, and differentiators.

By following this workflow, PMMs can use ChatGPT to create a clear, adaptable, and impactful messaging framework that drives alignment across teams and resonates with target audiences. Let me know if you'd like refinements!

Technical-to-Business Value Translation

For those doing more technical marketing as a PMM, turning technical capabilities into business outcomes that resonate with buyers can be quite time-consuming. After listening to countless sales calls, I've noticed decision-makers glaze over during technical discussions but lean in when the talk is about ROI and business impact.

Define the inputs:

- {industry=Insert your target industry, e.g., enterprise data infrastructure.}
- {solution=Insert your product/solution, e.g., automated data observability platform.}
- {audience=Insert your target audience, e.g., CFOs, IT leaders, procurement teams.}
- {technical features=Insert key technical features, e.g., anomaly detection algorithms, automated root cause analysis.}

Define the objective

You are a product marketing expert translating technical product features into clear, outcome-driven business value for {solution} in {industry}.
</role>
I need to communicate how the technical features of {solution} solve real-world business problems and deliver measurable outcomes for {audience}. This translation must avoid jargon and focus on business impacts, such as revenue growth, cost reduction, efficiency improvements, or risk mitigation.
</context>

Translate the following technical features into business value: [Insert a list of technical features, e.g., "automated anomaly detection" or "AI-based performance optimization"]. For each feature:
• Explain the capability in simple terms.
• Describe the business problem it addresses.
• Articulate the business outcome or value delivered (e.g., savings, efficiency, reduced risk).
</task>
<format>
Provide responses in a table format:

Technical Feature	Simplified Capability	Business Problem	Business Outcome
Feature 1: [Technical term]	[Simplified explanation]	[Challenge it solves for the audience]	[Outcome, e.g., 20% reduction in downtime]
Feature 2: [Technical term]	[Simplified explanation]	[Challenge it solves for the audience]	[Outcome, e.g., improved productivity]

</format>
<tone>
Clear, benefit-driven, and business-oriented.
</tone>

Simplify technical features

You are a technical communication expert skilled at simplifying complex features into plain language for business stakeholders.
</role>
I have a list of technical product features for {solution} that need to be explained in simple terms. Focus on making the capabilities easy to understand without technical jargon.

</context>

Rewrite the following technical features in simplified terms that a business executive can understand: [Insert technical features here].

</task>

<format>

Provide the response as a list:

- Feature 1: [Simplified explanation in one sentence].
- Feature 2: [Simplified explanation in one sentence].

Example:

- Technical feature: "AI-based anomaly detection."
  o Simplified Explanation: Automatically identifies unusual activity in real time to prevent costly disruptions.

</format>

<tone>

Clear, concise, and non-technical.

</tone>

Align features to business problems

You are a business strategist mapping technical features to real-world business problems for {solution}.

</role>

I need to demonstrate how the technical capabilities of {solution} address specific challenges faced by {audience}. Align each feature with a clear business problem it solves.

</context>

For each technical feature, identify:

- The **business problem** or challenge it resolves.
- The **pain point** this problem creates for the target audience.

</task>

Provide responses in bullet points:
- Feature: [Technical feature].
    o Business Problem: [The challenge it addresses].
    o Pain Point: [The impact or cost of not solving this problem].

Example:
- Feature: Automated anomaly detection.
    o Business problem: Systems go offline unexpectedly, causing downtime.
    o Pain point: Downtime results in lost revenue and damaged customer trust.
</format>
<tone>
Focused, problem-driven, and audience-relevant.
</tone>

Translate to measurable business outcomes

You are a value engineer quantifying the business impact of technical features for {solution}.
</role>
I need to articulate the measurable business outcomes delivered by the technical capabilities of {solution}. Focus on metrics that matter to business decision-makers, such as cost savings, efficiency gains, or risk reduction.
</context>
For each feature and its associated business problem, describe the business outcome using quantifiable results where possible. </task>
<format>
Provide the response as follows:
- Feature: [Technical feature].
    o Business outcome: [Quantifiable result, e.g., "20%

faster incident resolution reduces downtime by 30 hours per month"].

Example:

- Feature: Automated anomaly detection.
    - o Business outcome: "Reduces downtime by 20%, saving $50,000 per month in lost productivity."

</format>

<tone>

Quantitative, results-focused, and compelling.

</tone>

Create audience-specific business messaging

You are a product marketer tailoring business value messaging for {audience}. </role>

I need audience-specific messaging that translates the technical capabilities of {solution} into clear, compelling business value. Focus on addressing pain points, priorities, and goals unique to each audience segment.

</context>

For each technical feature:

Tailor the **value message** to resonate with the priorities of {audience}.

Highlight the **benefit or outcome** that aligns with their goals.

</task>

<format>

Provide responses in a structured format for each audience:

- Audience segment: [e.g., CFOs].
    - o Feature: [Technical feature].
    - o Value message: [Simplified benefit statement].
    - o Outcome: [Business result or KPI, e.g., "Reduce operational costs by 15% annually"].

Example:

- Audience segment: CFOs

o Feature: Automated anomaly detection.
o Value message: "Eliminate costly downtime by identifying system issues before they escalate."
o Outcome: "Save $500,000 annually by avoiding unplanned outages."

</format>

<tone>

Audience-specific, benefit-driven, and outcome-focused.

</tone>

Summarize the technical-to-business value framework

You are a senior strategist summarizing the business value of {solution} for executive stakeholders.

</role>

Summarize the translation of technical capabilities into business value, highlighting how {solution} solves real-world challenges and delivers measurable outcomes for {audience}.

</context>

Summarize the framework as follows:

1. Technical features simplified: A list of technical features explained in simple terms.
2. Business problems addressed: Key challenges the features resolve.
3. Business outcomes: Quantifiable results or benefits delivered.
4. Audience-specific messaging: Tailored value statements for key audience segments.

</task>

<format>

Provide the response as a structured summary with clear sections.

1. Technical features simplified:

o Feature 1: [Simplified explanation].
o Feature 2: [Simplified explanation].
2. Business problems addressed:
o Problem 1: [Description].
o Problem 2: [Description].
3. Business outcomes:
o Outcome 1: [Quantifiable result].
o Outcome 2: [Quantifiable result].
4. Audience-specific messaging:
o Segment 1: [Tailored message and outcome].
o Segment 2: [Tailored message and outcome].
</format>

<tone>
Executive-level, clear, and results-oriented.
</tone>

Why this workflow works

- **Simplifies complexity**: Breaks down technical features into clear, jargon-free language.
- **Aligns with business goals**: Connects product capabilities to outcomes like cost savings, efficiency, and risk reduction.
- **Audience relevance**: Tailors messaging to resonate with specific decision-makers and stakeholders.
- **Actionable outputs**: Provides structured deliverables that can be used directly in sales decks, marketing materials, and executive conversations.

By following this workflow, PMMs can use ChatGPT to translate technical features into clear, measurable business value that resonates with decision-makers and drives product adoption.

Practice Exercises

- **Develop a positioning framework**: For your product or a hypothetical one, use the provided prompts to define the target industry, solution, audience, and key competitors. Then, step through the prompts to identify your target customer and their main problem, define your product category and unique value, analyze competitors, and craft a final positioning statement. Share the framework with your team and discuss how it could inform go-to-market strategies.

- **Create a messaging framework**: Begin by specifying the target industry, solution, audience, and key value drivers. Follow the prompts to identify the target audience and their pain points, develop your core value proposition, create key messages with supporting proof points, and tailor messaging for different audience segments. Summarize the final messaging framework in a document that can be shared with sales, marketing, and customer success teams.

- **Practice translating technical features into business outcomes**: Choose three to five technical capabilities of your product or a hypothetical one. For each one, simplify the feature explanation, align it with a business problem it solves, quantify the business outcome it delivers, and tailor the value message for a specific audience like IT leaders or CFOs. Summarize the technical-to-business value framework in a presentation for executive stakeholders, highlighting how your product delivers real, measurable results.

Summary

- **Positioning framework creation**: Creating a positioning framework using Geoffrey Moore's Positioning Template—which includes defining the target customer, their problem, the product category, unique value proposition, and competitive differentiation—is a framework that helps crystallize a product's positioning in the market.

- **Messaging framework development**: A messaging framework that articulates a product's value to specific audiences involves identifying the target audience and their pain points, crafting a core value proposition, creating key messages with supporting proof points, and tailoring the messaging for different audience segments.

- **Technical-to-business value translation**: For those in technical marketing roles, translating technical product features into clear, outcome-driven business value is a must.

Chapter 5 References

[1] Antin, Alan, Michael Maziarka, and Molly Beams. "Positioning Revisited." Gartner. September 15, 2020. https://www.gartner.com/en/documents/3990177.

CHAPTER 7

Content and Communications

Content Strategy Framework

Okay, let's talk content strategy. As PMMs, we know that content is king. But it's not enough to just churn out random pieces of content. We need a plan, an actual strategy that aligns with our goals and resonates with our target audience.

Define inputs:

- {industry=Insert your target industry, e.g., enterprise cloud security.}
- {solution=Insert your product/solution, e.g., AI-driven threat detection platform.}
- {audience=Insert your target audience, e.g., CISOs, DevOps teams, IT security leaders.}
- {goals=Insert your content goals, e.g., lead generation, brand awareness, customer education.}

Define the objective

You are a content strategy expert specializing in {industry} and {solution}.
</role>

I need to create a content strategy that aligns with our business goals and resonates with {audience}. The strategy must focus on solving customer pain points, communicating product value, and guiding buyers through the decision-making process.

</context>

Develop a content strategy framework that includes:

- **Goals**: Define the primary content objectives (e.g., lead generation, customer education).
- **Audience**: Identify target segments and their needs.
- **Content themes**: Develop overarching themes tied to customer pain points and product messaging.
- **Content types**: Recommend formats (e.g., blogs, whitepapers, webinars) based on audience preferences.
- **Content calendar**: Suggest high-level content ideas mapped to the buyer journey stages.

</task>

<format>

Provide the response in a structured format:

1. Goals:
   - Goal 1: [Example: Increase inbound leads by 20% in 6 months].
   - Goal 2: [Example: Educate customers on best practices for cloud security].
2. Audience:
   - Segment 1: [role, e.g., "CISOs at enterprise companies"].
     - Needs: [What they need from content, e.g., "Risk management strategies"].
   - Segment 2: [role, e.g., "IT Managers at mid-market organizations"].
     - Needs: [e.g., "Simplified guides for secure cloud configurations"].

3. Content themes:
- o Theme 1: [Example: "Improving threat visibility and reducing response times"].
- o Theme 2: [Example: "Navigating compliance challenges in multi-cloud environments"].

4. Content types:
- o Top-of-funnel (awareness): Blog posts, social media snippets, short videos.
- o Middle-of-funnel (consideration): Case studies, webinars, whitepapers.
- o Bottom-of-funnel (decision): ROI calculators, comparison guides, product demos.

5. Content calendar ideas:
- o Awareness stage: [Example: Blog: "Top 5 security threats in multi-cloud setups"].
- o Consideration stage: [Example: Webinar: "How to achieve real-time threat detection"].
- o Decision stage: [Example: ROI Guide: "Save 30% on cloud monitoring costs with automation"].

</format>
<tone>
Strategic, structured, and actionable.
</tone>

Identify audience segments and content needs

You are a customer insights specialist analyzing content needs for {audience} in {industry}.
</role>
I need to understand the content preferences and pain points of our target audience to ensure the strategy addresses their needs effectively. Focus on how different roles consume content and what topics they care about.
</context>

For each target audience segment:
- Identify their **key challenges** or goals related to {solution}.
- Determine their **content preferences** (formats, channels).
- Suggest content topics or angles that will resonate with their needs.

</task>

<format>

Provide the response as follows:
- Audience segment 1: [role/title, e.g., CISOs].
  - Challenges: [Top 2–3 challenges, e.g., "Meeting compliance standards"].
  - Content preferences: [Formats: e.g., whitepapers, reports. Channels: LinkedIn, webinars].
  - Content ideas:
    - Topic 1: [E.g., "How to prepare for evolving cloud compliance regulations"].
    - Topic 2: [E.g., "Case study: Reducing threat response time in large organizations"].
- Audience segment 2: [role/title].
  - Challenges: […].
  - Content preferences: […].
  - Content ideas: […].

</format>

<tone>

Audience-focused, clear, and practical.

</tone>

Develop content themes and messaging

You are a messaging strategist defining key content themes and messages for {solution}.

</role>

\<context/\>

To ensure content aligns with our product positioning and audience needs, I need to develop content themes and supporting messages that highlight the value of {solution}.

\</context\>

\<task/\>

Develop 3–5 overarching content themes and create supporting messages for each theme. Align messages to customer pain points and business outcomes.

\</task\>

\<format\>

Provide the response as follows:

- Theme 1: [E.g., "Improving operational efficiency with automated threat detection"].
  - o Message 1: [How {solution} reduces manual effort for IT teams].
  - o Message 2: [Example: "Automated insights free up 20% of IT resources for strategic projects"].
- Theme 2: [E.g., "Navigating complex compliance requirements"].
  - o Message 1: [How {solution} ensures real-time compliance monitoring].
  - o Message 2: [Example: "Simplify audits with automated reporting and alerts"].

\</format\>

\<tone\>

Clear, benefit-driven, and consistent.

\</tone\>

Map content to the buyer journey

\<role/\>

You are a content strategist mapping content ideas to each stage of the buyer journey for {solution}.

\</role\>

I need to ensure that our content addresses customer needs at every stage of the funnel, from awareness to decision-making. Align content types and topics to buyer journey stages.
</context>
Map content ideas to the following stages:
- **Awareness stage**: Content that educates and builds interest.
- **Consideration stage**: Content that demonstrates value and differentiators.
- **Decision stage**: Content that supports purchasing decisions.
</task>
<format> Provide the response as a table:

Stage	Content Type	Content Idea
Awareness Stage	Blog, Social Post	[Example: "5 Cloud Security Risks You Must Know"]
Consideration Stage	Webinar, Case Study	[Example: "Case Study: Reducing Security Incidents by 50%"]
Decision Stage	ROI Guide, Demo Video	[Example: "ROI Calculator: Save on Cloud Threat Detection"]

</format>
<tone>
Structured, funnel-focused, and actionable.
</tone>

Create the content calendar outline

You are a content planner developing a high-level content calendar for {solution}. </role>

I need a quarterly content calendar outline that includes key content pieces, their objectives, and target audience segments.
</context>
Outline a 3-month content calendar that includes:
- Content title or topic.
- Objective (e.g., educate, generate leads, nurture).
- Target audience segment.
- Suggested content type.
</task>
<format> Provide the response in table format:

Month	Content Title/Topic	Objective	Audience	Content Type
Month 1	"Top 5 Trends in Cloud Security"	Build Awareness	IT Security Leaders	Blog Post
Month 2	"Reducing Threat Response Time"	Educate & Nurture	DevOps Teams	Webinar
Month 3	"ROI Calculator: Cost Savings"	Support Decision	CFOs	ROI Tool

</format>
<tone>
Practical, organized, and goal-driven.
</tone>

Why this workflow works

- **End-to-end framework**: Covers goals, audience needs, messaging, journey mapping, and calendar creation.

- **Audience-centric**: Focuses on solving customer problems at each stage of the funnel.
- **Structured outputs**: Delivers actionable, ready-to-implement deliverables like themes, content types, and calendar ideas.
- **Strategic alignment**: Ensures content aligns with business goals, audience priorities, and product messaging.

By following this workflow, PMMs can use GPT to create a scalable, results-driven content strategy framework that drives engagement, educates customers, and supports business growth.

The goal is to create a content engine that consistently delivers value to your audience and drives business results. AI can help build that engine and keep it running smoothly.

Blogs, eBooks, and Whitepapers

These are the heavy hitters of our content arsenal, providing in-depth analysis and establishing us as thought leaders in our space. But let's face it, finding the uninterrupted time to write a compelling piece of content can be challenging. These marquee content pieces require extensive research, careful structuring, and a clear, authoritative voice.

AI can help structure the content, generate initial drafts, and even suggest relevant data and research to support arguments.

Think about it. Instead of staring at a blank page, you can give the AI a topic and some key themes, which it can use to generate a detailed outline, complete with an executive summary, section headings, and even bullet points for key takeaways.

Example prompt

Variables (information needed):

- {industry = Insert the target industry, e.g., enterprise cloud security.}
- {solution = Insert the product/solution name, e.g., AI-driven threat detection platform.}
- {audience = Specify target roles and personas, e.g., CISOs, IT managers, compliance officers.}

- {content type = Define the type of content, e.g., whitepaper, eBook, blog.}
- {goal = Define the whitepaper's purpose, e.g., lead generation, thought leadership, customer education.}
- {topic = Central theme or challenge to address, e.g., "Reducing security risks with real-time anomaly detection".}
- {pain points = Audience pain points the whitepaper will solve, e.g., "Manual incident monitoring leads to delays and missed threats".}
- {business outcomes = Key outcomes or benefits delivered, e.g., "50% faster threat resolution and reduced downtime".}
- {data sources = Include supporting data or insights (internal research, industry stats, case studies).}

Define the objective and scope

You are a content strategist specializing in creating {content type} for {industry}.
</role>
I need to create a {content type} that educates {audience} on {topic}, addresses their core pain points, and positions {solution} as a leader in solving this challenge. The whitepaper should align with our business goal: {goal}.
</context>
Clarify the purpose, audience, and focus of the {content type}:
- **What is the primary objective**? (e.g., thought leadership, generating leads, or educating customers).
- **Who is the target audience**? (roles, industries, and pain points).
- **What specific challenge or trend will this whitepaper address**?
- **How does {solution} align with this topic and deliver value**?
</task>

Provide the responses as follows:

1. Objective: [e.g., "Establish thought leadership and generate leads"].

2. Target audience: [e.g., CISOs, IT leaders in financial services].

3. Key challenge: [e.g., "Manual anomaly detection leads to costly delays in resolving threats"].

4. Solution alignment: [e.g., "Our AI-driven platform automates anomaly detection to reduce resolution time by 50%"].

</format>

<tone>

Clear, goal-driven, and audience-focused.

</tone>

Develop a content outline

You are a content editor creating a structured {content type} outline to engage {audience} and align with {goal}.

</role>

The {content type} will focus on {topic} and address the pain points of {audience}. I need a clear outline that introduces the challenge, presents supporting evidence, and positions {solution} as the answer.

</context>

Create a {content type} outline using the following structure:

1. Title: Craft a clear, engaging title (e.g., "The Future of Threat Detection: How AI Reduces Security Risks").

2. Executive summary: Summarize the challenge, key insights, and the value delivered by {solution}.

3. Introduction: Frame the problem and its importance for {audience}.

4. The challenge: Explore the core pain points and business impact (e.g., manual workflows, missed threats).
5. Industry trends or data: Highlight trends, stats, or research that support the importance of solving the challenge.
6. Solution overview: Explain how {solution} solves the challenge and delivers measurable outcomes.
7. Proof points: Add supporting evidence, such as case studies, ROI data, or customer testimonials.
8. Conclusion/call-to-action: Summarize the value and provide a clear next step (e.g., demo request, contact).

</task>
<format> Provide the outline as follows:

1. Title: [{content type} title].
2. Executive summary: [2–3 sentences summarizing the whitepaper].
3. Introduction: [Brief framing of the problem].
4. The challenge:
   o Pain point 1: [Description].
   o Pain point 2: [Description].
5. Industry trends/data: [Relevant data points or trends].
6. Solution overview: [How {solution} addresses the challenge].
7. Proof points:
   o Case study: [Brief details of success story].
   o ROI data: [Quantifiable benefit].
8. Conclusion/CTA: [Summarize the solution and next steps].

</format>
<tone>
Structured, informative, and results-focused.
</tone>

Generate content for each section

You are a professional writer developing clear, engaging content for a {content type} aimed at {audience} in {industry}.

</role>

I need to draft the content for each section of the {content type} based on the approved outline. The tone must be professional, authoritative, and focused on delivering business value.

</context>

Draft the following sections of the {content type}:

1. The challenge: Clearly articulate the audience's key pain points, with real-world examples.

2. Industry trends: Integrate relevant stats, research, or benchmarks to validate the challenge.

3. Solution overview: Explain how {solution} addresses the problem in simple, outcome-driven language.

4. Proof points: Write a short customer success story or provide ROI-driven data to support claims.

</task>

<format>

Provide content drafts for each section:

1. The challenge:

   o [Content explaining the audience pain points, e.g., "Manual security processes result in delays that cost organizations $X annually"].

2. Industry trends:

   o [Content integrating supporting data, e.g., "According to Gartner, organizations adopting AI-driven security see 30% faster incident responses"].

3. Solution overview:

   o [Content explaining how {solution} solves the challenge, e.g., "Our AI-powered platform automates anomaly detection to minimize manual effort and reduce downtime"].

4. Proof points:

    o [Case study, testimonial, or ROI data, e.g., "Customer X reduced incident resolution time by 50%, saving $500K annually"].

\</format\>

\<tone\>

Authoritative, benefit-driven, and easy to follow.

\</tone\>

Refine content for tone and flow

\<role/\>

You are an editor refining the {content type} to ensure it flows logically, maintains professional tone, and drives engagement.

\</role\>

\<context/\>

I have drafted the {content type} content, and now I need you to refine it for clarity, consistency, and audience relevance.

\</context\>

\<task/\>

Review and improve the following content:

- Ensure clear transitions between sections.
- Simplify complex ideas and maintain a focus on outcomes.
- Optimize the language for {audience} to ensure readability and engagement.

\</task\>

\<format\>

Provide the refined content with suggested edits for tone, flow, and clarity. \</format\> \<tone\>

Professional, concise, and engaging.

\</tone\>

Finalize with visuals and a call to action

\<role/\>

You are a content designer advising on final touches to improve whitepaper engagement.

\</role\>

To ensure the {content type} is impactful, I need recommendations for visuals and a strong call to action.

</context>

Suggest:

- **Visuals**: Charts, infographics, or tables to reinforce key points.
- **Call to action**: A compelling next step that drives action (e.g., request a demo, download a resource).

</task>

<format>

Provide recommendations as follows:

1. Visuals:
   - Section: [e.g., "Industry Trends"].
   - Visual: [e.g., "Bar chart showing 30% incident response improvement"].
2. Call to action:
   - [Clear CTA, e.g., "Contact us for a personalized ROI analysis"].

</format>

<tone>

Actionable and visually focused.

</tone>

Why this workflow works

- **Structured process**: Guides PMMs from planning to step-by-step execution.
- **Audience alignment**: Ensures content directly addresses target audience pain points and goals.
- **Outcome-driven**: Focuses on business outcomes, proof points, and measurable value.
- **Engagement-ready**: Combines content clarity with strong visuals and a compelling CTA.

By following this workflow, PMMs can use GPT to create content that educates, engages, and drives measurable outcomes.

But don't just stop at the outline. Use AI to generate initial drafts of each section. This can give you a solid foundation to build upon. Then, collaborate with subject-matter experts to add depth, refine the arguments, and ensure technical accuracy.

The goal is to create whitepapers that are informative, persuasive, and establish thought leadership. AI can help by providing structure, generating content, and streamlining the writing process.

Pro tip:

After generating the initial content outline, ask the AI to go through each section and pause after each one to ask you questions on how it might be modified. This allows you to add your unique POV to the outline.

Here's a prompt that works quite well:

"Let's step through each section of the outline. Please suggest changes and pause after each section. Ask me questions for feedback. I will provide feedback and you can make the changes."

Email Sequence Design

Email. It's still a core tool for us PMMs, whether to nurture leads, promote new features, or reengage inactive users. But crafting effective email sequences that get results can be tricky. You need compelling subject lines, clear calls to action, and a message that resonates with an audience.

Variables (information needed):

- {industry = Target industry, e.g., enterprise cloud security.}
- {solution = Product name, e.g., AI-driven threat detection platform.}
- {whitepaper title = Title of the whitepaper, e.g., "The Future of Threat Detection: Reducing Security Risks with AI".}
- {audience = Target roles/personas, e.g., CISOs, IT security managers, DevOps teams.}

- {pain points = Key challenges the audience faces, e.g., manual monitoring, delayed incident responses, compliance stress.}
- {value proposition = Core value delivered, e.g., "Automates anomaly detection to reduce downtime and minimize risk."}
- {benefits = Tangible outcomes, e.g., 50% faster response time, improved visibility, cost savings.}
- {CTA = The call to action, e.g., "Download the whitepaper," "Learn more," "Request a demo".}

Create the landing page copy

You are a digital copywriter creating a high-converting landing page to promote the whitepaper: *{whitepaper title}*.
</role>

The whitepaper targets {audience} and addresses their key challenges, such as {pain points}. The goal is to encourage downloads by clearly communicating the value of the content and positioning {solution} as a thought leader.
</context>

Write landing page copy with the following components:
- **Headline**: Attention-grabbing statement highlighting the problem or value of the whitepaper.
- **Sub-headline**: Support the headline with a brief statement about what readers will learn.
- **Value bullets**: List 3–4 key takeaways or benefits of downloading the whitepaper.
- **Call to action (CTA)**: A clear, action-driven CTA.
</task>

<format> Provide the response in this structure:
- Headline: [Short, bold statement focused on audience pain point or solution].
- Sub-headline: [Supporting sentence describing the whitepaper's relevance].

- Value bullets:
  - Learn how to [benefit 1].
  - Discover strategies to [benefit 2].
  - Improve [specific outcome tied to audience pain point].
- CTA: [E.g., "Download the whitepaper to get started"].

Example Output:

- Headline: "Reduce Threat Response Time by 50% with AI-Driven Security Insights"
- Sub-headline: "Discover how automation and AI can help security teams minimize risks and streamline workflows."
- Value bullets:
  - Learn how to automate anomaly detection to catch threats early.
  - Discover real-world strategies to reduce incident response time.
  - Explore how leading enterprises save time and money with AI-powered tools.
- CTA: "Download the Whitepaper".

</format>

<tone>

Clear, benefit-driven, and actionable.

</tone>

Write the email nurture sequence

You are an email marketer crafting a 3-email nurture sequence to drive whitepaper downloads and engagement for {solution} in {industry}.

</role>

The whitepaper, titled *{whitepaper title}*, highlights {pain points} and demonstrates how {solution} delivers {benefits}. The audience includes {audience}, and the goal is to guide them to download the whitepaper and consider {solution}.

</context>

Write 3 nurture emails with the following structure:
- **Email 1: Awareness**: Introduce the whitepaper, highlight the audience's challenges, and offer the whitepaper as a solution.
- **Email 2: Value**: Reinforce the value of the content by showcasing benefits and key takeaways.
- **Email 3: Urgency/CTA**: Add a time-sensitive reason to download (e.g., exclusive insights, limited availability).
</task>
<format>
Provide the response as follows:
- Email 1: Awareness
  - Subject line: [E.g., "Struggling with security delays? Read this now"].
  - Body:
    - Highlight the challenge: [Audience pain point].
    - Introduce the whitepaper: [Title and relevance].
    - CTA: [E.g., "Download the whitepaper to solve this challenge"].
- Email 2: Value
  - Subject line: [E.g., "3 ways to reduce security risks—fast"].
  - Body:
    - Focus on benefits: [Key takeaways from the whitepaper].
    - Reiterate value: [What they will gain].
    - CTA: [E.g., "Get actionable insights today"].
- Email 3: Urgency
  - Subject line: [E.g., "Don't miss these security insights"].
  - Body:
    - Add urgency: [Why they should act now].
    - Highlight outcomes: [Specific benefits from reading the whitepaper].
    - CTA: [E.g., "Download now to stay ahead of security risks"].
</format>

<tone>
Concise, persuasive, and audience-focused.
</tone>

Generate social media promotions

You are a social media strategist creating promotional content to drive engagement and downloads for {whitepaper title}.
</role>

The goal is to promote the whitepaper across LinkedIn, Twitter, and Facebook to reach {audience}. Focus on highlighting {pain points} and the benefits of {solution}.
</context>

Write 3 promotional posts tailored to:
- **LinkedIn**: Professional and detailed, with a focus on audience pain points.
- **Twitter**: Short, punchy posts with a focus on key benefits.
- **Facebook**: Conversational and engaging, with a clear CTA.

</task>

<format>
Provide the response as follows:
- LinkedIn post:
- "Security teams spend hours chasing false positives and responding to threats too late. Our new whitepaper, {whitepaper title}, shows how AI-driven solutions reduce response times by 50% and improve risk management.
  Download it here: [CTA link]"
- Twitter posts:
    o "Manual threat detection = lost time. Learn how AI reduces response times by 50% in our latest whitepaper.
    Download now: [CTA link]"

> o "Cut through the noise. Discover actionable strategies to detect threats faster. Get the whitepaper: [CTA link]"
- Facebook post:
- "How much time does your team spend chasing alerts? Our latest whitepaper, {whitepaper title}, reveals how AI-driven tools reduce delays, save money, and improve security.
- Download now: [CTA link]."

</format>

<tone>

Platform-specific, benefit-driven, and engaging.

</tone>

Write digital ad copy

You are a digital advertising expert creating ad copy to promote the whitepaper, *{whitepaper title}*.

</role>

The target audience is {audience}, and the goal is to drive whitepaper downloads through digital ads (Google, LinkedIn, and Facebook). Focus on the core challenge {pain points} and the value delivered by {solution}.

</context>

Write ad copy for:
- **Google Ads**: Short headlines and descriptions.
- **LinkedIn Ads**: Professional, benefit-driven copy with a strong CTA.
- **Facebook Ads**: Engaging, clear copy that highlights benefits and a CTA. </task>

<format>

Provide the response as follows:
- Google Ads:
  - Headline 1: "Reduce Security Delays with AI"
  - Headline 2: "Download Our Security Whitepaper"

    o Description: "Discover how to reduce threat response times by 50%. Get actionable insights in our free whitepaper."
- LinkedIn Ads:
- "Manual security workflows lead to missed threats and delays. Learn how AI-driven tools are revolutionizing threat detection in our new whitepaper, {whitepaper title}.
- 📥 Download now: [CTA link]."
- Facebook Ads:
- "Tired of chasing alerts? Discover how AI helps security teams detect and respond faster. Download our free whitepaper today: [CTA link]."

</format>

<tone>

Clear, concise, and conversion-focused.

</tone>

Why this workflow works

- **Unified messaging**: Ensures consistency across all campaign touchpoints.
- **Platform optimization**: Tailors content for different channels while maintaining alignment.
- **Audience focus**: Highlights pain points and benefits that resonate with the target audience.
- **Clear CTAs**: Drives conversions with strong, action-oriented messaging.

By following this workflow, PMMs can use GPT to efficiently generate campaign copy that promotes a whitepaper across multiple channels, driving engagement and measurable results.

Practice Exercises

- **Content strategy challenge**: Select a hypothetical B2B product in a specific industry. Using the framework outlined in this chapter, develop a comprehensive content strategy that maps out audience segments, content themes, and a quarterly content calendar. Demonstrate how the strategy addresses specific customer pain points and aligns with business goals.

- **Whitepaper development workshop**: Choose a complex industry challenge and create a full whitepaper outline using the step-by-step workflow in this chapter. Include a draft of the landing page copy, develop a potential email nurture sequence, and craft social media posts to promote the whitepaper. Focus on creating a cohesive narrative that showcases thought leadership and drives audience engagement.

- **Integrated campaign simulation**: Design a multichannel marketing campaign for a B2B solution, incorporating the techniques discussed in the chapter. Create a video script, develop digital ad copy, and draft a series of social media posts. Ensure the campaign tells a consistent story across different platforms, highlighting the product's value proposition and addressing specific audience pain points.

Summary

- **AI-powered content strategy**: AI can assist in generating comprehensive content plans, including diverse formats like blog posts, whitepapers, social media updates, and customer stories.

- **Streamlined case study development**: AI can leverage customer quotes, interview transcripts, and survey data to generate compelling case study narratives tailored to specific audiences.

- **Efficient blog post and whitepaper creation**: AI can generate initial drafts and outlines for blog posts and whitepapers, saving time and effort while allowing for human refinement and expertise.

CHAPTER 8

Launch and Campaign Planning

Launch Planning

Let's talk product launches, the culmination of months of hard work, the moment we unveil our latest creation to the world. But a successful launch requires meticulous planning and coordination across multiple teams.

That's where a comprehensive launch checklist comes in. It helps us stay organized, track progress, and ensure we haven't overlooked any critical steps.

But creating these checklists can be tedious. There are so many details to consider, from marketing and sales enablement to technical readiness and legal compliance.

AI can be a real lifesaver.

Define inputs:

- {product name= The name of the product being launched, e.g., "AI-driven Threat Detection Platform".}
- {industry=<insert industry> = Target industry, e.g., enterprise cloud security.
- {audience}: Target personas, e.g., CISOs, IT leaders, DevOps teams.}
- {launch goal = Primary objective, e.g., lead generation, customer acquisition, driving product adoption, or increasing market awareness.}

- {key value proposition = The product's main value, e.g., "Reduces threat response time by 50% with automated anomaly detection."}
- {key features/benefits = List 3–5 major features and their outcomes.
- {competitors = Main competitors, e.g., Competitor A, Competitor B.}
- {launch date = Planned date for launch.}
- {launch assets = Materials needed, e.g., landing page, whitepaper, social campaigns, emails, videos.}

Define the launch goals and success metrics

You are a product marketing strategist planning a product launch for {product name} in {industry}.
</role>
I need a clear definition of the launch goals and measurable success metrics. The plan must align with the needs of {audience} and the overall business objectives.
</context>
Answer the following questions to define the goals:
- **What is the primary goal** of this launch? (e.g., lead generation, market awareness, customer acquisition).
- **What are the key success metrics**? (e.g., number of signups, leads generated, revenue targets, demo requests).
- **What challenges or competitive threats** might impact the success of this launch?
</task>
<format>
Provide the responses as follows:
1. Launch goal: [E.g., Generate 1,000 qualified leads in the first month].
2. Success metrics:
    o Metric 1: [E.g., "500 demo requests within 6 weeks"].

o Metric 2: [E.g., "10% conversion rate from email nurture campaigns"].

3. Challenges/threats: [E.g., "Competitor A has a similar feature launching next quarter"].

</format>

<tone>

Strategic, focused, and goal-driven.

</tone>

Develop the product positioning and messaging

You are a messaging strategist defining the positioning and key messages for {product name}.

</role>

To ensure a successful product launch, I need clear positioning, a strong value proposition, and messaging that resonates with {audience}. This messaging must highlight {key features/benefits} while differentiating from {competitors}.

</context>

Develop the following:

• **Value proposition**: A concise statement summarizing how {product name} solves {audience}'s challenges.

• **Key messages**: 3–5 benefit-driven messages addressing the audience's pain points.

• **Differentiation**: How {product name} stands out from {competitors}.

</task>

<format>

Provide the response in this format:

1. Value proposition: "[Product name] helps [audience] solve [key problem] by [unique benefit]."

2. Key messages:
   - o Message 1: [E.g., "Reduce manual threat detection efforts with automated anomaly detection."]
   - o Message 2: [E.g., "Improve incident response time by 50%, minimizing downtime."]
   - o Message 3: [E.g., "Simplify compliance workflows with real-time reporting tools."]
3. Differentiation:
   - o Competitor A: [Strength/weakness].
   - o Our differentiator: [What makes us unique, e.g., "AI-driven root cause analysis with 90% accuracy"].

</format>
<tone>
Clear, benefit-driven, and differentiator-focused.
</tone>

Plan the launch timeline and key activities

You are a launch planner creating a detailed timeline and activity plan for the product launch of {product name}.
</role>
The launch date is {launch date}, and I need a clear schedule of activities leading up to and following the launch. Activities must include messaging, asset creation, internal enablement, and external promotion.
</context>
Create a launch timeline divided into:
- **Pre-launch activities**: Preparation, messaging, and asset development.
- **Launch day activities**: Campaign activation and announcements.
- **Post-launch activities**: Continued promotion and performance tracking. </task>
<format>

Provide the response in a table format:

Phase	Timeline	Activities	Owner
Pre-Launch	[e.g., 4 weeks before]	Develop messaging, create landing page, finalize emails.	PMM Team
Launch Day	[Date]	Activate campaigns, publish social posts, send launch email.	Marketing Team
Post-Launch	[e.g., 4 weeks after]	Analyze campaign results, host a webinar, nurture leads.	Demand Gen Team

</format>
<tone>
Organized, actionable, and structured.
</tone>

Develop launch content and assets

You are a content strategist creating key launch assets to drive awareness, engagement, and conversions for {product name}.
</role>
I need to develop content and assets that align with the product's messaging, resonate with {audience}, and achieve {launch goal}.
</context>
List the essential content assets for the product launch, including:

1. **Landing page**: Headline, sub-headline, and value bullets.
2. **Email sequence**: 3-email nurture sequence (awareness, value, CTA).

3. **Social media posts**: Copy for LinkedIn, Twitter, and Facebook.

4. **Ad copy**: Short-form digital ad content for Google and LinkedIn.

5. **Sales enablement**: A one-pager or battlecard to equip internal teams.

</task>

<format>

Provide responses as follows:

1. Landing page:
   o Headline: [e.g., "Transform Security Operations with Automated Threat Detection"].
   o Sub-headline: [e.g., "Reduce response times and minimize downtime with AI-driven insights."].
   o Value bullets:
     ▪ Automates threat detection and resolution.
     ▪ Reduces false positives by 50%.
     ▪ Saves teams hundreds of hours annually.

2. Email sequence:
   o Email 1: Awareness—[Brief description and CTA].
   o Email 2: Value—[Highlight benefits and outcomes].
   o Email 3: Urgency—[Why they should act now and CTA].

3. Social media posts:
   o LinkedIn: ["Introducing [product name]! Automate security operations and reduce response times by 50%. Discover more: [link]"].
   o Twitter: ["Tired of slow security response? Meet [product name]—AI-driven insights to act faster. 🚀 Learn more: [link]"].

4. Ad copy:
   o Google Ads:
     ▪ Headline: "Cut Threat Response Time by 50%."
     ▪ Description: "Discover how [product name] automates detection to minimize downtime. Learn more now."

5. Sales enablement:
   o Key differentiators: [List benefits].
   o Customer proof: [Data/statistics].
</format>
<tone>
Clear, persuasive, and aligned with messaging.
</tone>

Monitor and optimize post-launch performance

You are a performance analyst tracking the success of {product name}'s launch.
</role>
I need to monitor key metrics post-launch, evaluate performance, and suggest optimization opportunities to maximize impact.
</context>
List the metrics to track and provide recommendations for optimization:
* **Metrics**: Identify KPIs (e.g., landing page conversion rate, email open rates, demo requests).
* **Optimization recommendations**: Suggestions for improving campaign performance.
</task>
<format>
Provide the response as follows:
1. Metrics:
   o Landing page conversion rate: [Target: 10%].
   o Email open rate: [Target: 25%].
   o Demo requests: [Target: 200 in 6 weeks].
2. Optimization recommendations:
   o Refine CTAs to improve landing page conversions.
   o Test different subject lines for email engagement.
   o Increase ad spend on high-performing social platforms.

</format>

<tone>

Data-driven, results-focused, and actionable.

</tone>

Why this workflow works

- **Comprehensive planning**: Covers strategy, content, execution, and measurement.
- **Audience-centric**: Aligns messaging and assets with audience pain points and needs.
- **Clear deliverables**: Ensures all required content is outlined and developed.

PR and Announcement Strategy

Press releases. They're how we get the word out to the media, generate buzz, and position our product in the market. But crafting one that gets attention and tells a compelling story can be intricate.

Variables (information needed):
- {announcement type=Define the type of announcement, e.g., product launch, funding round, major partnership.}
- {product or initiative=The product, solution, or initiative being announced, e.g., AI-powered threat detection platform.}
- {audience=Target stakeholders, e.g., customers, investors, analysts, journalists.}
- {key message=The core message you want to communicate, e.g., "Reduces security response time by 50% with AI automation."}
- {key benefits=Tangible outcomes or benefits of the announcement, e.g., faster performance, cost reduction, improved efficiency.}
- {spokesperson=The individual(s) quoted in the announcement, e.g., CEO, CMO.}
- {proof points=Supporting facts, stats, or customer quotes, e.g., "Customer X achieved a 30% reduction in downtime."}

- {call to action (CTA)=The next step for the audience, e.g., "Learn more on our website," "Request a demo," or "Contact us for interviews."}

Define the announcement goals and audience

You are a PR strategist planning a media announcement for {announcement type} regarding {product or initiative}.
</role>
To ensure the announcement achieves its objectives, we need to define the goals and the audience it will target. The announcement must communicate a clear, compelling story that resonates with key stakeholders.
</context>
Answer the following questions:
- **What is the primary goal of the announcement**? (e.g., media coverage, investor interest, product adoption, brand awareness).
- **Who is the target audience**? (e.g., customers, analysts, journalists, investors).
- **What message or outcome do we want the audience to take away**?
</task>
<format>
Provide the response as follows:
1. Goal: [E.g., "Generate press coverage and drive 500 website visits within 24 hours."]
2. Target audience: [E.g., "Tech journalists, IT decision-makers, and industry analysts."]
3. Key message: [E.g., "{Product} revolutionizes anomaly detection by reducing security response time by 50%."]
</format>

Strategic, clear, and outcome-focused.

</tone>

Draft the press release

You are a PR writer crafting a professional press release to announce {announcement type} for {product or initiative}.

</role>

The press release should follow a clear structure, focus on the core message, and include quotes, proof points, and a strong call to action.

</context>

Write a press release with the following structure:

1. **Headline**: A strong, concise headline that grabs attention.
2. **Sub-headline**: Supporting text to expand on the headline.
3. **Lead paragraph**: State the announcement clearly—who, what, when, where, and why it matters.
4. **Body**:
   o Describe the problem the product/initiative solves.
   o Highlight key features, benefits, and proof points.
   o Include a quote from a company spokesperson.
5. **Conclusion**: Summarize the announcement and provide a clear CTA.

</task>

<format>

Provide the press release as follows:

1. Headline: "[Product] Launches to Revolutionize [Benefit/Outcome] for [Audience]."
2. Sub-headline: "[Product] delivers [key result] through [unique approach]."

3. Lead paragraph:

"[Company Name] today announced [product or initiative], a solution designed to [key benefit]. This launch marks a milestone in [industry challenge or trend]."

4. Body:
- o The problem: [State the audience challenge].
- o The solution: "[Product] automates [process] to deliver [key outcome]."
- o Key features/benefits:
  - Benefit 1: [E.g., "Reduces response times by 50%."]
  - Benefit 2: [E.g., "Eliminates manual workflows with AI automation."]
- o Quote: "[Spokesperson Name], [Title], says: '[Insightful quote reinforcing the value of the announcement].'"
- o Proof points: [Stat, customer result, or differentiator, e.g., "Customer X reduced downtime by 30%."]

5. Conclusion:

"To learn more about [product/initiative], visit [link] or contact [press contact]."

</format>

<tone>

Professional, concise, and newsworthy.

</tone>

Plan media outreach

You are a PR strategist building a media outreach plan to maximize coverage for {announcement type}.

</role>

I need a list of media targets, outreach messaging, and a pitch outline to engage journalists, analysts, and relevant media outlets.

</context>

Develop a media outreach plan that includes:

1. **Media list**: Key journalists, publications, or influencers to target.
2. **Email pitch**: A concise, engaging pitch email to secure media interest.
3. **Follow-up plan**: Timeline and strategy for follow-up communication.

</task>
<format>

Provide responses as follows:

1. Media list:
   - o Publication 1: [e.g., TechCrunch]
   - o Journalist 1: [Name, focus area]
   - o Publication 2: [e.g., VentureBeat]
2. Email pitch:
   - o Subject line: [E.g., "New AI Platform Cuts Security Response Time in Half"].
   - o Body:
   - o "Hi [Journalist Name],
   - o I hope you're doing well. I'm reaching out to share an exclusive on our latest launch: [Product/Initiative].
   - o Why it matters: [Brief description of the problem it solves and the result it delivers].
   - o Would you be interested in an interview or early access to learn more?
   - o Let me know if we can share more details!
   - o Best,
   - o [Your Name, Contact Information]"
3. Follow-up plan:
   - o Initial pitch: [Date]
   - o Follow-up email: [2 days after pitch].
   - o Additional call/reminder: [One week post-launch].

</format>
<tone>

Concise, engaging, and tailored to the media.

</tone>

Develop supporting launch assets

You are a content strategist creating supporting materials for the PR announcement of {product or initiative}.
</role>
To amplify the announcement, we need additional assets that media outlets, analysts, and customers can engage with.
</context>
List and describe supporting materials, such as:
1. **Landing page**: Summarize the announcement with a CTA.
2. **Product one-pager**: Highlight features and benefits.
3. **Social media posts**: Promotional posts for LinkedIn and Twitter.
4. **Visuals**: Infographics, screenshots, or short videos.
</task>
<format>
Provide responses as follows:
1. Landing page:
    o Headline: [Reinforce the key message].
    o CTA: [E.g., "Learn More"].
2. Product one-pager:
    o Overview: [Problem, solution, and benefits].
    o Features: [List 3 key features].
3. Social media posts:
    o LinkedIn: "[Product] has arrived! Automate [process] and improve [outcome] by 50%. Learn more: [link]."
    o Twitter: "We're excited to announce [Product]—your answer to [pain point]! 🐦 See it here: [link]."
4. Visuals:
    o Infographic: [Show how {product} solves the problem visually].
    o Screenshots: [Highlight the product in action].
</format>

Cohesive, engaging, and actionable.

</tone>

Measure results and optimize

You are a PR analyst evaluating the impact of the PR announcement for {product or initiative}.

</role>

I need to measure the success of the announcement and identify opportunities for improvement.

</context>

List the key performance metrics and recommend optimization strategies.

</task>

<format>

1. **Key metrics**: Media coverage: [Number of articles published]; Website traffic: [Increase in visits to the landing page]; CTA performance: [Clicks, demo requests, or downloads].

2. Optimization recommendations:
   o Message refinement: [Improve headlines based on media feedback].
   o Outreach: [Expand media list or adjust timing].
   o Content: [Enhance visuals or add new proof points].

</format>

<tone>

Data-driven, clear, and actionable.

</tone>

Why this workflow works

- **Strategic foundation**: Aligns messaging and goals with audience priorities.
- **Clear structure**: Provides a repeatable format for announcements.

- **Multichannel focus**: Combines press, content, and social strategies for maximum impact.
- **Measurable results**: Tracks performance to optimize future efforts.

By following this workflow, PMMs can use GPT to craft and execute a comprehensive PR announcement strategy that drives awareness, engagement, and action.

Pro tip:

Use Perplexity.ai to gather competitive insights. This will help position your announcement effectively in the market and highlight what makes your product stand out.

The goal is to generate media coverage that reaches a target audience and drives interest in your product. AI can generate press releases that are newsworthy, informative, and engaging.

Practice Exercises

- **Launch narrative challenge:** Develop a compelling launch narrative for a new AI-powered cybersecurity platform targeting enterprise customers in regulated industries. Craft a clear value proposition, 3-5 key messaging points, and a concise product description that resonates with decision-makers.
- **Launch planning workshop:** Create a comprehensive launch plan for a major update to a cloud-based data management platform. Develop a timeline with key milestones, a launch checklist covering various functions, and a framework for measuring success metrics.
- **PR strategy simulation:** Design a PR strategy for a significant product announcement in your industry. Draft a press release, plan media outreach, and create supporting assets like a landing page and social media posts. Outline how you'll measure the impact of the announcement.

Summary

- **Launch checklists ensure thorough planning**: AI can generate comprehensive launch checklists that cover pre-launch, launch day, and post-launch activities and help PMMs stay organized and on track.

- **Communication sequences orchestrate messaging**: AI can assist in designing effective communication sequences across various channels, ensuring consistent messaging and audience engagement.

- **Presentations, PR, and success metrics are key**: AI can streamline the creation of launch presentations, press releases, and success metric frameworks, saving time and effort while maintaining quality.

CHAPTER 9

Sales and Partner Enablement

Sales Play Generation

Time to move on to sales enablement. As PMMs, we know that our sales team is on the front lines; they need the right tools and resources to be successful. That's where sales plays come in. They provide a structured framework for engaging with prospects, handling objections, and closing deals.

Variables (information needed):

- {product name = Name of the solution, e.g., AI-driven threat detection platform.}
- {industry = Target industry, e.g., financial services, enterprise cloud infrastructure.}
- {audience = Buyer personas, e.g., CISOs, IT security leaders, procurement managers.}
- {pain points = Core challenges faced by the audience, e.g., manual monitoring, slow resolution, compliance issues.}
- {value proposition = Clear value statement, e.g., "Reduce response times by 50% with automated detection."}
- {key differentiators = What sets the product apart from competitors, e.g., "AI-driven root cause analysis with 90% accuracy."}

- {sales scenario = Specific sales situation, e.g., competitive deal, new product launch, expansion opportunity.}
- {resources needed = Sales enablement materials, e.g., customer stories, ROI calculators, product demos, battlecards.}
- {CTA = The action the sales rep should drive, e.g., "Schedule a demo," "Book a workshop," "Start a pilot."}

Define the sales play objective and scenario

You are a sales enablement strategist creating a sales play for {product name} in {industry}.

</role>

The purpose of this sales play is to help sales teams engage {audience} in {sales scenario}. The play will highlight {pain points}, position {product name} as the solution, and provide the tools to move opportunities forward.

</context>

Define the key components of the sales play:

1. **Objective**: What is the goal of this sales play? (e.g., win competitive deals, drive adoption, reengage inactive accounts).
2. **Sales scenario**: Describe the situation this play addresses (e.g., objection handling, competitive positioning).
3. **Target audience**: Who is the play for? What role, industry, and challenges are they facing?

</task>

<format>

Provide responses as follows:

1. Objective: [E.g., "Win competitive deals against Competitor A by highlighting our AI automation benefits."]

2. Sales scenario: [E.g., "Deals where customers are evaluating manual vs. automated anomaly detection tools."]

3. Target audience:
   o Role: [E.g., "CISOs and IT security leaders."]
   o Challenges: [E.g., "Manual workflows delay threat detection and increase risk."]

</format>

<tone>

Strategic, focused, and clear.

</tone>

Develop the sales play structure

You are a sales strategist designing the step-by-step structure for {sales scenario} to enable sales reps to position {product name} effectively.

</role>

The sales play should guide the rep through:

- The opening hook (how to engage the customer).
- Discovery questions (to uncover pain points).
- Key talking points (highlighting {value proposition} and {key differentiators}).
- Customer proof (case studies or ROI examples).
- The call to action (next step to drive).

</context>

Outline the sales play structure as follows:

1. **Opening hook**: An attention-grabbing statement to engage the buyer.

2. **Discovery questions**: 3–5 questions to uncover pain points and needs.

3. **Key talking points**: 3–5 messages highlighting {product name}'s value.

4. **Customer proof**: Add a relevant case study, stat, or result.

5. **CTA**: The action sales reps should aim for in the conversation.

</task>

<format>

Provide the sales play structure as follows:

1. Opening hook:

   "How much time is your team spending chasing false security alerts? What if you could reduce response times by 50%?"

2. Discovery questions:

   o "How are you currently detecting and responding to security threats?"

   o "What impact are false positives or manual workflows having on your team's productivity?"

   o "Are there specific compliance or downtime concerns you're facing?"

3. Key talking points:

   o Automation advantage: "{Product name} automates anomaly detection, reducing manual effort and response delays."

   o Faster incident resolution: "Our AI reduces response times by 50%, minimizing downtime and risk."

   o Accuracy: "Unlike competitors, our platform reduces false positives by 60%, saving hours of unnecessary work."

4. Customer proof:

   o "Customer X cut incident resolution time by 50%, saving $200K annually."

5. CTA:

   "Let's schedule a demo to show you how {product name} can solve these challenges in your environment."

</format>

<tone>

Actionable, benefit-driven, and easy to follow.

</tone>

Create supporting sales assets

You are a sales enablement expert developing materials to support the sales play for {product name}.
</role>
To help sales reps execute this play effectively, we need supporting content, such as battlecards, ROI calculators, and product demos.
</context>
List and describe the sales assets needed to support this sales play, including their purpose and usage.
</task>
<format>
Provide responses as follows:

1. Sales battlecard:
   - o Purpose: Equip sales reps with key talking points, customer objections, and competitive differentiators.
   - o Usage: Quick reference during calls to handle objections and position {product name}.
2. ROI calculator:
   - o Purpose: Quantify the value of {product name} for the buyer.
   - o Usage: Use during the decision stage to show cost savings and efficiency gains.
3. Case study:
   - o Purpose: Provide real-world proof of success.
   - o Usage: Share with buyers who need validation from peers or industry examples.
4. Product demo:
   - o Purpose: Showcase key capabilities and benefits in action.
   - o Usage: Conduct live demos tailored to the buyer's pain points.

</format>

<tone>
Practical, clear, and results-focused.
</tone>

Write an email template to share the sales play

You are a sales enablement leader sharing the new sales play with the sales team to help them engage customers effectively.
</role>
The email should introduce the play, explain its purpose, and highlight key elements like opening hooks, talking points, and assets.
</context>
Write an internal email to share the sales play with the sales team.
</task>
<format>
Provide the email as follows:
- Subject line:
- "New Sales Play: Win Competitive Deals with [Product Name]"
- Body:
- "Hi Team,
- We've created a new sales play to help you close competitive deals and highlight the unique value of [Product Name].
- Why it matters:
- Our customers face challenges like [key pain points], and this play equips you to position [Product Name] as the best solution.
- What's included:
  o Opening hook: Grab attention with customer pain points.
  o Discovery questions: Uncover their challenges and needs.

- o Key talking points: Position our value and key differentiators.
- o Supporting assets: Battlecards, ROI calculators, and case studies.
- Action:
- Review the attached play and use it in your next customer call. Reach out if you need help preparing!
- [CTA: Link to the sales play resource]

Best,

[Your Name]

Sales Enablement Team"</format>

<tone>

Clear, motivating, and action-oriented.

</tone>

Measure and optimize sales play performance

You are a sales performance analyst evaluating the impact of the {sales play} for {product name}.

</role>

I need to track how well the sales play is working and identify areas for improvement.

</context>

List the key performance metrics and recommend optimization strategies. </task> <format>

Provide responses as follows:

1. Key metrics:
   - o Usage rate: How often the play is being used.
   - o Conversion rate: Deals won using this play.
   - o Feedback: Sales rep and customer responses.
2. Optimization recommendations:
   - o Refine discovery questions based on customer conversations.

o Add new proof points if customers request more validation.

o Update the CTA based on deal stage feedback.

</format>

<tone>

Data-driven, actionable, and focused on continuous improvement.

</tone>

Why this workflow works

- **Actionable steps**: Provides a clear structure for building and executing a sales play.
- **Audience alignment**: Tailors messaging and questions to customer pain points and goals.
- **Resource support**: Equips sales teams with the tools to succeed.
- **Measurable impact**: Tracks performance for ongoing refinement.

By following this workflow, PMMs can use GPT to create a structured, repeatable sales play that empowers their sales teams to close deals efficiently and consistently.

The goal is to equip the sales team with the tools they need to succeed. AI can create comprehensive sales plays tailored to the target market and aligned with overall sales strategy.

Sales Cheat Sheet

Sales reps are busy people—juggling calls, meetings, and demos—all while trying to keep up with the latest product updates and competitive intel. They need information that's quick and easy to access, something they can refer to in the heat of the moment.

That's where a good cheat sheet comes in. It's a concise summary of key product information, competitive advantages, and common objections, all distilled into a handy one-pager.

But creating one can be a pain. It takes time to gather the information, organize it logically, and make sure it's easy to digest.

With AI, you can use this kind of prompt:

<context>

Our sales team needs a quick reference guide for our data analytics platform.

</context>

<task>

Create a cheat sheet that includes:

- A concise product overview.
- Key features and benefits.
- Target customer profile.
- Top 3 competitive advantages.
- Common objections and how to address them.
- Pricing and packaging options.

</task>

<format>

Provide the information in a bullet-point format that is easy to scan and digest.

</format>

Pro tip:

Use visuals to make the cheat sheet even more effective. Include a product screenshot, a competitive comparison chart, or even a simple flowchart that illustrates the sales process.

The goal is to empower the sales team with the information they need—when they need it. AI can produce concise, easy-to-reference cheat sheets that are always at their fingertips.

Battlecard Development

Battlecards are a sales rep's secret weapon in the competitive arena. A good battlecard provides a quick and easy way to make comparisons with a competitor's product, highlight strengths, and exploit weaknesses.

But creating effective battlecards can be a challenge. You need to gather competitive intel, analyze product features, and distill it all into a concise, easy-to-digest format.

AI can help.

Variables (information needed):

- {product name = Name of your product/solution, e.g., AI-driven Threat Detection Platform.}
- {industry = Target industry, e.g., financial services, cloud security, enterprise SaaS.}
- {audience = Buyer personas, e.g., CISOs, IT managers, DevOps leaders.}
- {key value proposition = A concise product value statement, e.g., "Reduces incident response time by 50% through automated anomaly detection."}
- {competitor name(s) = Competitors to compare against, e.g., Competitor A, Competitor B.}
- {competitor strengths = Key advantages of the competitor, e.g., "Established market presence, lower cost for SMBs."}
- {competitor weaknesses = Gaps or shortcomings, e.g., "Manual processes, limited scalability, high false positives."}
- {differentiators = Your product's unique strengths, e.g., "AI-driven automation, real-time insights, seamless scalability."}
- {objections and responses = Common objections sales reps encounter and responses to overcome them.}
- {proof points = Supporting customer quotes, data, or ROI examples, e.g., "Customer X reduced downtime by 30% and saved $200K annually."}

Define the battlecard objective and audience

You are a sales enablement strategist creating a competitive battlecard for {product name} in {industry}.

</role>

The battlecard will help sales reps position {product name} effectively against {competitor name(s)}. It must be concise, actionable, and focused on differentiating {product name} for {audience} while addressing competitor strengths and weaknesses.

</context>

Define the following components of the battlecard:

1. **Target audience**: Who will use the battlecard and for which buyer personas?
2. **Objective**: What is the purpose of the battlecard? (e.g., competitive positioning, objection handling, value articulation).

</task>

<format>

Provide responses as follows:

1. Target audience:
   o Sales team: [E.g., "Account Executives and SDRs engaging CISOs and IT leaders."]
   o Buyer personas: [E.g., "CISOs facing incident response challenges in large enterprises."]
2. Objective:
   o [E.g., "Equip sales reps to position {product name} against {competitor name}, highlight key differentiators, and address customer objections."]

</format>

<tone>

Clear, strategic, and actionable.

</tone>

Develop competitor comparison

You are a competitive intelligence analyst creating a comparison of {product name} against {competitor name(s)} for a battlecard.

</role>

The battlecard must include a competitive overview, highlighting competitor strengths, weaknesses, and {product name}'s differentiators. The goal is to equip the sales team with concise talking points to handle comparisons effectively and position {product name} as the superior choice.

</context>

Provide a competitive comparison as follows:

- **Competitor overview**: Strengths and weaknesses of {competitor name(s)}.
- **Differentiators**: Highlight how {product name} outperforms competitors.

</task>

<format>

Provide the responses in table format:

Criteria	{Competitor Name 1}	{Product Name}
Strengths	[Competitor strength 1, strength 2]	[Unique strength 1, strength 2]
Weaknesses	[Competitor gap 1, gap 2]	[No weaknesses listed here]
Key Differentiators	[N/A]	[Key differentiator 1, differentiator 2]

Example:

Criteria	Competitor A	Product Name
Strengths	Strong market presence; lower price	Faster implementation; better AI
Weaknesses	High false positives; manual workflows	Scalable automation; proven ROI
Key Differentiators	N/A	Reduces response times by 50%; Seamless integrations

</format>

Clear, concise, and competitor-focused.

</tone>

Key sales talking points

You are a sales strategist identifying talking points to help sales teams position {product name} effectively against {competitor name(s)}.

</role>

The sales team needs 3–5 concise, benefit-driven talking points that emphasize {product name}'s advantages and align with the needs of {audience}. Each point should address common customer pain points and how {product name} solves them.

</context>

List 3–5 key sales talking points that:

Highlight how {product name} solves specific pain points.

Address customer concerns where competitors fall short.

</task>

<format>

Provide the talking points as follows:

- Talking point 1: [E.g., "Automates anomaly detection, reducing manual effort by 70%."]
- Talking point 2: [E.g., "Delivers 50% faster incident response, minimizing costly downtime."]
- Talking point 3: [E.g., "Scales seamlessly across enterprise environments, unlike Competitor A."]
- Talking point 4: [E.g., "Reduces false positives by 60%, improving team efficiency."]
- Talking point 5: [E.g., "Simplifies compliance workflows with automated reporting tools."]

</format>

<tone>
Benefit-driven, persuasive, and customer-focused.
</tone>

Handle common objections

You are a sales enablement expert helping sales teams handle objections effectively for {product name}.
</role>
Sales reps often encounter objections when positioning {product name} against {competitor name(s)}. Create clear, confident objection responses that emphasize value and address concerns.
</context>
Identify 3 common objections and provide responses for each, using real proof points or product differentiators.
</task>
<format>
Provide responses as follows:
- Objection 1: "[Customer says, 'Competitor A is cheaper.']"
  - Response: "While Competitor A may have a lower upfront cost, {product name} delivers greater long-term ROI by automating processes and reducing incident response times by 50%. For example, Customer X saved $200K annually after switching to our platform."
- Objection 2: "[Customer says, 'We're comfortable with our current solution.']"
  - Response: "Many of our customers felt the same before realizing how much time and money they were losing with manual workflows. With {product name}, teams save 20+ hours weekly and achieve faster, more accurate threat resolution."
- Objection 3: "[Customer says, 'We don't have time for a lengthy deployment.']"

o Response: "{Product name} is designed for rapid deployment—most of our customers go live within two weeks, minimizing disruption while delivering immediate value."

\</format>

\<tone>

Confident, empathetic, and results-oriented.

\</tone>

Include customer proof points

\

You are a product marketer providing real-world proof points to strengthen the sales battlecard for {product name}.

\</role>

\

Proof points build credibility and help sales reps validate claims when engaging with {audience}. Include success metrics, case studies, or customer quotes.

\</context>

\

List 2–3 customer proof points to include in the battlecard.

\</task>

\<format>

Provide proof points as follows:

- Proof point 1: "Customer X reduced incident response time by 50%, saving $200K annually."
- Proof point 2: "Organizations using {product name} reduced false positives by 60%, improving team productivity."
- Proof point 3: "Customer Y scaled their security operations across 10 global regions seamlessly with {product name}."

\</format>

\<tone>

Data-driven, concise, and credible.

\</tone>

Summarize the battlecard

<role/>

You are a product marketing manager finalizing the battlecard summary for {product name}.

</role>

<context/> The final battlecard should include:

- **Competitive overview**
- **Key talking points**
- **Objection handling**
- **Proof points**

</context>

<task/>

Summarize the battlecard into one cohesive document for sales teams to use during customer conversations.

</task>

<format>

Provide the battlecard summary as follows:

1. Competitive overview:
   o Competitor strengths: [Strength 1, Strength 2].
   o Competitor weaknesses: [Weakness 1, Weakness 2].
   o Our differentiators: [Unique value points].
2. Key talking points:
   o Point 1: [Benefit-driven message].
   o Point 2: [Benefit-driven message].
3. Objection handling:
   o Objection: "[Customer concern]."
     ■ Response: "[Reassuring response with proof point]."
   4.Proof points:
   o "Customer X reduced downtime by 30%."
   o "Customer Y scaled to 10 global regions seamlessly."

</format>

<tone>

Concise, structured, and sales-ready.

</tone>

Why this workflow works

- **Competitive clarity**: Provides a clear comparison to help sales reps differentiate effectively.
- **Benefit-driven messaging**: Equips sales teams with impactful talking points that resonate with customer pain points.
- **Objection handling**: Prepares sales reps to address common objections confidently.
- **Proof points**: Builds credibility through real-world examples and results.
- **Actionable format**: Structured for quick reference during live conversations.

By following this workflow, PMMs can use GPT to create a battlecard that empowers sales teams to handle competitive scenarios, articulate value, and close deals more effectively.

The goal is to give a sales team the ammunition needed to win competitive deals. AI can yield concise, easy-to-reference battlecards that arm teams with the right information at the right time.

Practice Exercises

- **Sales play development challenge**: Select a hypothetical B2B product in a specific industry and develop a comprehensive sales play from scratch. Create the entire sales enablement package, including the sales play structure, supporting assets (battlecard, ROI calculator, email template), and a measurement framework. Focus on crafting a sales approach that addresses specific customer pain points and provides a clear, actionable path for the sales team to engage and convert prospects.

- **Competitive battlecard workshop**: Choose two competing products in a technology market and create a detailed battlecard that comprehensively compares their strengths, weaknesses, and unique value propositions. Develop a set of talking points, objection-handling strategies, and customer proof points that position the hypothetical product as the superior solution. Demonstrate the ability to translate competitive intelligence into a powerful sales enablement tool.

- **Sales enablement content simulation**: Design a complete sales enablement content suite for a complex B2B solution, including:
  - A sales cheat sheet with key product information.
  - A comprehensive battlecard against a key competitor.
  - An internal email introducing the sales play to the team.
  - A set of discovery questions and key talking points.
  - Potential objection responses with supporting proof points.

Summary

- **AI-powered sales playbooks**: AI can assist in generating comprehensive sales playbooks, including tailored messaging, objection handling scripts, and competitive positioning for different buyer personas.
- **Streamlined sales cheat sheets**: AI can efficiently create concise and easily digestible sales cheat sheets summarizing key product information, competitive advantages, and objection-handling strategies.
- **Automated battlecard and comparison matrix generation**: AI can leverage product information and competitive intelligence to generate detailed battlecards and comparison matrices, aiding sales teams in competitive positioning.

Related Disciplines and Applications

CHAPTER 10

Technical Product Marketing

Feature Announcements

This is where we get to spread the word about all the cool new things we're building. But let's be honest, writing compelling announcements that resonate with different audiences can be tough. You've got to strike the right balance between technical detail and clarity and tailor the message for different channels.

Instead of writing separate announcements for an email list, blog, and press release, you can use AI to generate variations that hit the right tone for each audience.

Example prompt

<context>
We're launching a new feature that significantly improves the scalability of our data analytics platform. We want to highlight the cost reduction benefits for IT leaders.
</context>
<task>
Write a product feature announcement for IT leaders emphasizing scalability and cost reduction.
</task>

Provide three versions of the announcement:
1. A concise email announcement.
2. A detailed blog post.
3. A formal press release.
</format>

Pro tip:

Use AI tools that are specifically designed for generating marketing copy (like Jasper). They often have built-in templates and frameworks that can help create professional-looking announcements in minutes.

The goal is to get people excited about your new features. AI can help by quickly generating compelling announcements tailored to the target audience.

Technical Integration Guides

Integrations are essential these days. Customers need to seamlessly connect our platform with their existing tech stack, whether it's syncing data with Salesforce, leveraging the power of a cloud data warehouse like Snowflake, or running advanced analytics with Databricks. Clear, concise integration guides are crucial for making this happen.

But creating these guides can be a real headache. Accurate technical instructions need to be provided while ensuring they're also user-friendly, even for people not deep in the technical weeds. And the nuances of integrating with different platforms—from hyperscalers like AWS, Google Cloud, and Microsoft Azure to specialized services like Snowflake and Databricks—need to be considered.

AI can step in and make things easier. It can help develop those step-by-step instructions that are easy to follow, regardless of someone's technical background.

Imagine feeding the AI details about your platform and the integration target, and letting it generate a comprehensive guide.

Example prompt

<context>
We need to help customers integrate our platform with various services, including Salesforce, Snowflake, and Databricks. We also need to provide guidance on deploying our platform on AWS, Google Cloud, and Microsoft Azure.
</context>
<task>
Draft an integration guide for each of the following scenarios:
- Syncing our platform with Salesforce.
- Connecting our platform to Snowflake.
- Integrating our platform with Databricks.
- Deploying our platform on AWS.
- Deploying our platform on Google Cloud.
- Deploying our platform on Microsoft Azure.
</task>
<format>
For each scenario, provide a step-by-step guide with clear instructions, code samples where relevant, and screenshots.
</format>

To make such guides even better, we can upload customer FAQs or past integration feedback. This allows the AI to learn from previous challenges and provide more helpful solutions.

The goal is to make integrations smooth and painless. AI can make user-friendly guides that empower customers to connect our platform with the tools and services they rely on, whether it's a CRM giant like Salesforce, a data powerhouse like Snowflake, or the vast ecosystems of the major cloud providers.

Demo and Video Script Development

Video is a powerful medium for storytelling. Whether customer testimonials, product explainer videos, demos, or thought leadership pieces, video can capture attention and engagingly convey complex information. However, creating compelling video content requires a well-crafted script.

Variables (information needed):

- {industry = Target industry, e.g., enterprise cybersecurity.}
- {solution = Product or solution name, e.g., AI-driven threat detection platform.}
- {audience = Target roles/personas, e.g., CISOs, IT managers, DevOps leaders.}
- {pain points = Core challenges your audience faces, e.g., manual monitoring, false positives, delayed threat response.}
- {key value proposition = A clear, concise product value statement, e.g., "Automates anomaly detection to reduce downtime and minimize security risks."}
- {benefits = Tangible outcomes delivered, e.g., "50% faster response times, improved productivity, reduced costs." }
- {call to action (CTA) = The next step you want the viewer to take, e.g., "Request a demo," "Start a free trial," or "Download our whitepaper."}

Define the objective and format

You are a video marketing strategist specializing in crafting short, compelling video scripts to promote {solution} in the {industry}.

</role>

The goal is to create a 1–3 minute video script that engages {audience}, highlights their pain points, communicates the key benefits of {solution}, and encourages them to take action with a clear CTA.

</context>

Define the video structure and objective:

- **Objective**: What do we want the video to achieve? (e.g., awareness, lead generation, product explanation).

- **Tone**: What tone should the video have? (e.g., conversational, authoritative, inspiring).
- **Video format**: Choose the style of video: Animated explainer; Product demo with voiceover; Live-action customer story or talking head.

</task>
<format>
Provide responses as follows:

1. Objective: [e.g., "Generate awareness of {solution} and drive demo requests."]
2. Tone: [e.g., Conversational, clear, and confident.]
3. Video format: [e.g., Animated explainer video to simplify technical concepts.]

</format>
<tone>
Focused, clear, and strategic.
</tone>

Structure the video script outline

You are a video content strategist outlining a 1–3 minute promotional video for {solution}.
</role>
The script must capture attention, highlight the audience's challenges, showcase how {solution} solves these problems, and end with a strong call to action. The outline will follow this structure:

1. Hook: Grab attention in the first 10–15 seconds.
2. Problem statement: Clearly state the audience's key challenge.
3. Solution introduction: Introduce {solution} as the answer to the problem.
4. Benefits: Highlight 2–3 key benefits that matter most to the audience.

5. Proof point: Add a stat, customer result, or differentiator to build credibility.

6. Call to action (CTA): End with a clear next step.

</context>

Create a video script outline with the above structure for {solution}. </task> <format>

Provide the outline as follows:

1. Hook: [Opening line that grabs attention, e.g., "Security breaches cost millions—can you detect them before it's too late?"]

2. Problem statement: [Describe the audience's pain point].

3. Solution introduction: [Brief intro to {solution} and its purpose].

4. Benefits:

   a. Benefit 1: [Key value delivered, e.g., "Automates threat detection to save time."]

   b. Benefit 2: [E.g., "Reduces false positives, improving team productivity."]

5. Proof point: [Data, customer result, or unique differentiator, e.g., "Organizations using {solution} see a 50% faster response time."]

6. CTA: [Action you want the viewer to take, e.g., "Request a demo to see it in action."]

</format>

<tone>

Engaging, benefit-driven, and structured.

</tone>

Write the full video script

You are a video scriptwriter creating a 1–3 minute promotional video for {solution}.

</role>

I need a polished script that follows this flow:
- Hook (15 seconds)
- Problem statement (20–30 seconds)
- Solution introduction (20 seconds)
- Benefits (30–60 seconds)
- Proof point (15 seconds)
- Call to action (10–15 seconds)

The tone should be {tone}, and the content must speak to {audience} by addressing their pain points and showcasing the tangible outcomes of {solution}.
</context>

Write the full video script based on the outline provided.
</task>

<format>
Provide the script with timestamps as follows:

[0:00–0:15] Hook

"[Opening statement to grab attention.]"

[0:15–0:30] Problem statement

"[Describe the challenge your audience faces and why it matters.]"

[0:30–0:50] Solution introduction

"Introducing [solution]—[simple, clear statement of what it does and why it's the answer]."

[0:50–1:30] Benefits

"Here's how [solution] helps:
- Benefit 1: [e.g., "Automates anomaly detection to save your team hours every week."]
- Benefit 2: [e.g., "Reduces false positives so you focus only on real threats."]
- Benefit 3: [e.g., "Enables 50% faster threat response, reducing downtime and risks."]"

[1:30–1:45] Proof Point

"Organizations using [solution] have reduced their incident response times by 50%, saving $500,000 annually."

[1:45–2:00] Call to Action

"Want to see how it works? Request a demo today and take control of your security operations."

</format>

<tone>

Engaging, benefit-focused, and conversational.

</tone>

Optimize for delivery and visual suggestions

You are a creative director optimizing the video script for delivery and visuals.

</role>

The script needs visual suggestions (graphics, animations, or product shots) to reinforce key points and engage the audience effectively.

</context>

For each section of the script, suggest corresponding visuals or animations to enhance the content. </task> <format> Provide visual suggestions as follows:

- Hook: [Dynamic animation or text on screen: "Security breaches cost $4 million on average—are you prepared?"]
- Problem statement: [Show visuals of frustrated IT teams, security alerts flooding a dashboard.]
- Solution introduction: [Product logo and animated product overview with key features highlighted.]
- Benefits:
  - Benefit 1: [Animation of alerts being resolved automatically.]
  - Benefit 2: [Split screen showing "manual monitoring" vs. "automated results."]
  - Benefit 3: [Chart showing 50% improvement in response times.]

- Proof point: [Overlay a stat or testimonial: "Customer X cut downtime by 50%, saving $500K annually."]
- CTA: [Text on screen: "Request Your Demo Today" with a clickable button.]

</format>

<tone>

Creative, engaging, and visual-focused.

</tone>

Finalize with a review pass

You are a content editor ensuring the script flows smoothly, maintains clarity, and aligns with the desired tone.

</role>

Review the script for flow, audience resonance, and visual alignment to ensure the final output is polished and ready for production.

</context>

Provide feedback or final adjustments to improve clarity, engagement, and alignment with {solution}'s positioning.

</task>

<format>

Provide suggestions as follows:

1. Script adjustments: [Any refinements needed to tighten messaging].
2. Visual tweaks: [Suggestions for improving visuals or transitions].

</format>

<tone>

Polished, clear, and professional.

</tone>

Why this workflow works

- **Structured approach**: Step-by-step guide to creating an engaging, audience-focused video script.
- **Audience alignment**: Ensures messaging addresses audience pain points and outcomes.
- **Time-conscious**: Focuses on a clear, concise structure that fits the 1–3 minute format.
- **Visual integration**: Combines words with visuals to maximize impact.
- **Action-oriented**: Ends with a clear CTA to drive measurable results.

By following this workflow, PMMs can use GPT to efficiently create high-impact video scripts that resonate with their audience, communicate product value, and drive action.

The goal is to create videos that are informative, engaging, and drive results. AI can help do this by generating high-quality scripts that capture attention and communicate the value of our products or services.

Practice Exercises

- **Integration guide challenge**: Select three different technology platforms (e.g., a CRM, a cloud service, and a data analytics tool) and create comprehensive integration guides for a hypothetical product. Develop these guides using the AI-assisted workflow discussed in the chapter, ensuring the output is both technically accurate and accessible to users with varying levels of technical expertise. Include step-by-step instructions, code samples, potential troubleshooting tips, and visual aids.

- **Feature announcement multichannel campaign**: Develop an announcement for a complex technical product feature. Create variations of it for different channels (email, blog post, press release, social media) that maintain a consistent core message while adapting to the specific tone and requirements of each platform. Demonstrate how to translate technical details into compelling, audience-specific communications.

- **Technical content localization exercise**: Take a technical integration guide or feature announcement and demonstrate how to adapt it for three different audience segments (e.g., enterprise IT leaders, startup technical founders, and mid-market operations managers). Showcase how to modify the technical depth, language, and value proposition to resonate with each specific audience while maintaining the core technical accuracy of the content.

Summary

- **Technical precision meets accessibility**: Unlike the previous chapter's focus on broad content strategy, this chapter emphasizes the critical skill of translating complex technical features into clear, actionable content. Product marketing managers must bridge the gap between intricate technical details and user-friendly communication, making complex integrations and features understandable to diverse audiences.

- **Practical documentation as a strategic asset**: While Chapter 7 explored content creation as a marketing tool, Chapter 8 positions technical documentation and integration guides as crucial product assets. These materials are not just support content but strategic tools that can directly impact customer adoption, reduce support tickets, and showcase the product's technical sophistication and flexibility.

- **AI as a technical communication accelerator**: Whereas the previous chapter used AI for content generation and strategy, this chapter demonstrates its role in solving specific technical communication challenges. The focus shifts from creative content generation to precise, context-aware technical writing that can rapidly produce accurate, nuanced integration guides and feature explanations across multiple technical platforms and audience segments.

CHAPTER 11

Analyst Relations

Supposedly, analysts are the sages of the B2B world, the trusted advisors and oracles that buyers turn to for guidance. Whether you agree with that or not, their reports, rankings, and recommendations can make or break a product launch, influence purchasing decisions, and shape market perceptions so it's important to play the game.

Below are two workflows: 1) to understand analysts' perspectives, and 2) to help with the dreaded RFPs included in most analyst evaluations.

Analyst Relations Workflow

Variables (information needed):

- {product name = Name of the product/solution, e.g., AI-driven anomaly detection platform.}
- {industry = Target industry, e.g., cloud security, enterprise data governance.}
- {audience = Target analysts (firm and roles), e.g., Gartner, Forrester, IDC analysts specializing in security operations.}
- {analyst coverage areas = Key topics or market spaces analysts focus on, e.g., cloud security posture management, threat detection automation.}

- {company worldview = Your company's perspective or thesis about the industry, product category, and emerging trends.}
- {key differentiators = Unique product strengths, e.g., "AI-driven automation reduces false positives by 60%."}
- {analyst perspective = Analysts' reported views, e.g., "The market is prioritizing solutions with rapid ROI and easy integrations."}
- {AR goals = Primary objectives for AR, e.g., influence Magic Quadrant position, secure inclusion in analyst reports, align messaging for customer impact.}
- {proof points = Supporting data or case studies, e.g., "Customer X reduced incident resolution time by 50%."}

Understand the analyst's perspective

You are an AR strategist analyzing the perspective of {audience} to align {company worldview} with their market outlook.
</role>
To build strong analyst relationships and secure favorable coverage, we need to deeply understand the analysts' views on our market and how our positioning aligns with or challenges their perspective.
</context>
Generate insights on the analyst perspective by:
- Summarizing the latest analyst reports or thought leadership on {industry}.
- Highlighting the trends, priorities, and expectations analysts emphasize.
- Identifying any gaps or differences between {company worldview} and the analysts' perspective.
</task>
<format>
Provide responses as follows:
1. Analyst perspective summary:

o "[E.g., Analysts believe automation and real-time insights are critical priorities for reducing operational overhead in cloud security.]"

2. Key trends and priorities:
   o Trend 1: [E.g., "Rapid ROI is a deciding factor for enterprise buyers."]
   o Trend 2: [E.g., "Integrated solutions that work across hybrid environments are gaining traction."]

3. Perspective gaps:
   o "[Our company emphasizes AI accuracy, but analysts focus on deployment speed and ease of integration. This gap needs to be addressed in messaging.]"

</format>
<tone>
Objective, analytical, and insight-driven.
</tone>

Map your worldview to the analyst perspective

You are a strategic advisor aligning {company worldview} with the analyst perspective on {industry}.
</role>
We need to reconcile any differences between our perspective and analysts' views to present a cohesive, credible narrative that resonates with both analysts and customers.
</context>
Map your company's worldview to the analyst perspective by: Identifying where your worldview **aligns** with analyst priorities.

Highlighting where your perspective **challenges** current analyst thinking and why.

Creating bridging statements to connect your differentiators to the analysts' priorities.
</task>

Provide responses as follows:
1. Alignment areas:
   - "[E.g., Both analysts and our company agree that automation is the key to reducing operational overhead.]"
2. Challenging analyst views:
   - "[E.g., Analysts underemphasize the role of AI accuracy, whereas our solution demonstrates that accuracy directly drives efficiency and cost savings.]"
3. Bridging statements:
   - "[E.g., While speed of deployment is important, achieving 60% fewer false positives through AI automation ensures lasting ROI and reduced team workload.]"
</format>
<tone>
Balanced, strategic, and persuasive.
</tone>

Develop a clear AR narrative

You are an AR strategist crafting a compelling narrative for analysts to position {product name} as a leader in {industry}.
</role>
The narrative must:
- Address analyst priorities and align with their market perspective.
- Emphasize {product name}'s key differentiators and market impact.
- Highlight customer success stories and tangible proof points.
</context>
Draft the AR narrative by:

Creating a positioning statement for {product name}.

Outlining 3–4 key messages that connect product strengths to analyst priorities.

Integrating proof points to substantiate claims.

</task>

<format>

Provide the AR narrative as follows:

1. Positioning statement:
   o "[Product name] delivers automated [capability] that solves [industry challenge] for [audience] with [unique differentiator]."

2. Key messages:
   o Message 1: [E.g., "Our AI-driven automation reduces incident response times by 50%, aligning with the market need for efficiency and speed."]
   o Message 2: [E.g., "Seamless integrations across hybrid environments meet enterprises' growing demand for flexibility."]
   o Message 3: [E.g., "Customer success proves that AI accuracy translates directly into ROI, reducing false positives by 60%."]

3. Proof points:
   o "[Customer X reduced downtime by 30%, saving $200K annually.]"
   o "[Deployment time was cut to 2 weeks with seamless cloud integrations.]"

</format>

<tone>

Clear, benefit-driven, and aligned to analyst priorities.

</tone>

Prepare analyst briefing materials

You are an AR content strategist developing materials for analyst briefings on {product name}.

</role>

<context/>

We need to create a set of clear, concise materials to guide analyst briefings, ensuring they align with {analyst perspective} while communicating our company's narrative.

</context>

<task/>

Create the following briefing materials:

- **Analyst presentation outline**: Slide-by-slide breakdown of topics.
- **Supporting materials**: Case studies, proof points, and competitive insights. </task>

<format>

Provide responses as follows:

1. Analyst presentation outline:
   - Slide 1: Title Slide (Company and Product Overview).
   - Slide 2: Industry Trends (Analyst-Aligned Perspective).
   - Slide 3: The Problem We Solve (Customer Challenges).
   - Slide 4: Our Solution (Overview of {product name} and key differentiators).
   - Slide 5: Key Benefits (Mapped to analyst priorities).
   - Slide 6: Customer Proof Points (Case studies and ROI metrics).
   - Slide 7: Competitive Positioning (Our unique strengths).
   - Slide 8: Call to Action (Next steps for ongoing analyst engagement).

2. Supporting materials:
   - Customer case studies: [Provide 1–2 success stories].
   - Competitive overview: [Highlight 3 differentiators against key competitors].
   - ROI metrics: [Quantifiable proof of value].

</format>

<tone>

Organized, professional, and data-driven.

</tone>

Follow up and refine

<role/>

You are an AR analyst tracking feedback and refining the analyst relations strategy based on insights gained during briefings.

</role>

<context/>

After the analyst briefing, we need to:

- Capture analyst feedback and questions.
- Refine our narrative and materials based on insights.
- Plan for follow-up engagements and relationship building.

</context>

<task/>

Summarize analyst feedback and recommend next steps.

</task>

<format>

Provide responses as follows:

1. Key feedback:
    o "[E.g., Analysts were interested in deployment speed but needed more clarity on ROI impact.]"
2. Refinement opportunities:
    o "[E.g., Add customer data highlighting faster time-to-value.]"
3. Next steps:
    o "[E.g., Schedule a follow-up session to provide deeper insights on scalability and customer outcomes.]"

</format>

<tone>

Responsive, iterative, and action-oriented.

</tone>

Why this workflow works

- **Analyst alignment**: Deeply understands the priorities of analysts and maps positioning to their perspective.
- **Clear narrative**: Communicates product value with proof points that resonate with the research focus of analysts.

- **Actionable materials**: Provides structured, analyst-ready content to support impactful briefings.
- **Iterative refinement**: Captures feedback and adapts the strategy for stronger alignment and credibility.

By following this workflow, PMMs can use GPT to create a comprehensive analyst relations strategy, fostering analyst buy-in, influencing coverage, and aligning product narrative with market expectations.

Analysts can provide an objective perspective on products, competitors, and the market landscape. Their feedback can help refine messaging, identify blind spots, and ensure a product that resonates with buyers.

Executive Summary

Reports are essential for communicating insights and informing decisions. But let's be honest, not everyone has the time or inclination to read a 20-page market analysis or a 50-page competitive assessment.

That's where a concise and informative executive summary comes in. It provides a snapshot of the key findings, highlighting the most important takeaways and recommendations.

But writing effective executive summaries can be a challenge. We need to distill complex information into a concise and engaging narrative that captures the attention of busy executives.

AI can help generate those summaries, ensuring they're clear, concise, and tailored to our audience.

Example prompt

<context>
We have a 10-page market research report on the future of cloud computing.
</context>
<task>
Write a 2-paragraph executive summary of this 10-page market research report, focusing on:

- Key market trends and growth drivers.
- Emerging technologies and their potential impact.
- Competitive landscape and key players.
- Opportunities and challenges for our company.

</task>
<format>
Provide a summary that is concise, informative, and uses clear, non-technical language.
</format>

Pro tip:

Create different versions of the executive summary for different audiences. A summary for a CEO might focus on strategic implications and high-level trends, while a summary for a product manager might delve into more tactical details and product-specific recommendations.

The goal is to provide busy executives with the information they need to make informed decisions. AI can deliver that by generating concise and informative executive summaries that cut to the chase and highlight the most important takeaways.

RFPs for Analyst Evaluations

Like the Olympics of the B2B world, analyst evaluations are where products are put to the test and ranked against the competition. And the key to success in this arena is a well-crafted RFP (Request for Proposal).

Prerequisites

To effectively map technical capabilities to the RFP:

1. **Gartner RFP spreadsheet:**
   o **Full document containing**:
     - **Technical requirements**: Row-by-row requirements, features, specifications, or questions (e.g., scalability, integrations, performance metrics).

- **Columns for responses**: Blank fields to complete (e.g., Yes/No, feature description, proof points).
- **Priorities or weighting (if provided)**: Which requirements are critical, optional, or differentiators?

2. **Product documentation (technical details)**:
   - ○ **Detailed product technical information, including**:
     - **Key features**: Full feature list with descriptions.
     - **Technical specifications**: Deployment models, architecture, performance, compliance certifications, and scalability metrics.
     - **Integrations**: Supported APIs, environments, or third-party tools.
     - **Limitations**: Known constraints or feature gaps.

3. **Previous RFP Responses**:
   - ○ **Completed RFPs for similar evaluations, including**:
     - **Common answers**: Responses to frequently repeated questions.
     - **Proof points**: Benchmarks, customer metrics, or case studies used previously.
     - **Formatting**: Any required tone, structure, or template preferences.

Extract and summarize the RFP requirements

You are a technical analyst summarizing the key technical requirements from a **Gartner RFP spreadsheet** into organized themes.

</role>

The RFP contains 200+ rows of requirements, and we need to:

- Extract all technical requirements.
- Group them into high-level categories (e.g., **security, performance, scalability, integrations**).
- Flag any critical or weighted sections that require special attention.

</context>

Summarize the RFP requirements into organized categories to simplify mapping. </task>

<format>

Provide responses in the following structure:

1. High-level categories:
   o Category 1: [E.g., Security].
     ▪ Row 1: "Does the product comply with SOC 2 Type II standards?"
     ▪ Row 12: "Is end-to-end encryption supported?"
   o Category 2: [E.g., Performance].
     ▪ Row 33: "What is the data ingestion rate for real-time streams?"
     ▪ Row 45: "Does the solution support sub-5ms latency for 100K events?"
2. Critical sections:
   o Highlight critical rows or weighted requirements: [E.g., Rows 10–25 have a weighting of 30%].

</format>

<tone>

Clear, structured, and organized for further analysis.

</tone>

Map product technical details to RFP requirements

You are a technical content specialist mapping the product's technical capabilities to the RFP requirements. Use **product documentation** and **previous RFP responses** to ensure consistency and accuracy.

</role>

Each RFP row must be mapped to a clear, concise response that aligns with product capabilities. Include supporting details, proof points, and standard answers from prior RFPs where relevant.

</context>

For each technical requirement in the RFP, complete the following steps:

- Map the **requirement** to the corresponding product feature or specification.
- Add a clear, concise response describing the feature or technical capability.
- Include supporting proof points (benchmarks, certifications, or customer metrics).
- Flag any gaps or limitations that require further review.

</task>
<format>

Provide the response in table format:

RFP Row	Requirement	Product Capability	Response	Proof Point	Gaps/ Notes
12	"Does the product support SOC 2 Type II?"	Compliance Certifications	"Yes, {product name} is SOC 2 Type II certified."	Certification document available.	None
33	"What is the data ingestion rate?"	Data Ingestion Performance	"Supports ingestion of up to 10K events per second with 2ms latency."	Verified with Customer X benchmark.	None
45	"Does the product integrate with AWS?"	Cloud Integrations	"Yes, {product name} integrates natively with AWS S3, Redshift, and Lambda."	AWS Partnership Case Study.	None
150	"Can the solution operate on-premise?"	Deployment Models	"Supports both cloud and on-premise deployments."	Deployed successfully with Customer Y.	Requires technical verification.

</format>
<tone>
Concise, accurate, and professional.
</tone>

Address gaps and clarify limitations

You are a technical advisor identifying any gaps or limitations when mapping the product's technical capabilities to the Gartner RFP.
</role>
Some RFP requirements may not align perfectly with current product capabilities. We need to:

- Flag any feature gaps or partial support.
- Provide alternative solutions or explanations where possible.
- Clarify known limitations transparently to maintain credibility.

</context>
Identify rows where full support is not available and suggest alternative responses. </task>
<format>
Provide responses as follows:

1. Feature gaps:
   o Row: [E.g., 150].
   o Requirement: "Does the product offer native Kubernetes support?"
   o Current status: "Partial support through custom integrations."
   o Alternative response: "While native support is under development, {product name} integrates with Kubernetes clusters via standard APIs."
2. Limitations:
   o Row: [E.g., 175].
   o Requirement: "Supports 1M concurrent users."

- o Current status: "Scales up to 500K users per deployment."
- o Clarification: "Scaling beyond 500K concurrent users requires additional cluster configuration."

</format>

<tone>

Transparent, solution-focused, and professional.

</tone>

Incorporate responses from previous RFPs

You are a technical content manager ensuring consistency in responses by referencing previous RFPs completed for similar evaluations.

</role>

We want to maintain consistent language, proof points, and formatting for repeated RFP requirements.

</context>

For recurring requirements, retrieve standardized answers and align the current responses accordingly.

</task>

<format>

Provide the responses in table format:

RFP Row	Requirement	Standardized Response	Source RFP
12	"Does the product support SOC 2 Type II?"	"Yes, our solution is SOC 2 Type II compliant, certified annually."	RFP 2023 - Tech Company A

| 45 | "Integrations with AWS?" | "Native integrations with AWS services, including S3, EC2, and Lambda." | RFP 2022 - FinTech B |

```
</format>
<tone>
```
Consistent, standardized, and reliable.
```
</tone>
```

Final review and optimization

```
<role/>
```
You are a technical editor performing the final review of the completed RFP responses to ensure alignment, clarity, and technical accuracy.
```
</role>
<context/>
```
The RFP must be reviewed for completeness, consistency, and differentiation against competitors.
```
</context>
<task/>
```
Review all responses, validate technical accuracy, and suggest final optimizations to emphasize differentiators.
```
</task>
<format>
```
Provide review notes as follows:

1. Completeness: All rows addressed.
2. Consistency: Standardized responses for repeated questions.
3. Differentiation: Emphasize unique capabilities in responses (e.g., "Unlike Competitor X, we deliver 60% faster ingestion rates.").

```
</format>
```

<tone>
Meticulous, clear, and competitive.
</tone>

Why this workflow works

- **Handles scale**: Breaks down hundreds of RFP rows into manageable, organized steps.
- **Leverages inputs**: Combines the RFP document, product documentation, and prior responses for efficiency and consistency.
- **Ensures accuracy**: Maps capabilities while transparently addressing gaps or limitations.
- **Drives differentiation**: Highlights strengths and proof points to stand out in evaluations.

By following this workflow, PMMs and technical teams can use GPT to efficiently map product capabilities to Gartner RFP requirements while ensuring precision, completeness, and alignment.

Practice Exercises

- **Develop an analyst narrative**: Choose a B2B technology product and craft a compelling positioning narrative that aligns with current market analyst perspectives. Research recent analyst reports in the product's industry and create a strategic narrative that bridges the company's unique value proposition with the analysts' key priorities and market trends.

- **RFP response simulation**: Select a complex technical product and complete a mock Gartner RFP evaluation by mapping the product's capabilities to a comprehensive set of technical requirements. Develop responses that not only address each requirement accurately but also highlight the product's unique differentiators and provide concrete proof points.

- **Competitive positioning exercise**: Create a detailed analyst relations strategy that positions a product against key competitors in its market space. Develop a comprehensive approach that includes analyzing analyst perspectives, identifying unique product strengths, and creating a strategic plan for engaging and influencing key industry analysts.

Summary

Analyst relations are critical for B2B marketers. By understanding the role of analysts, crafting effective RFPs, and engaging with influencers strategically, their influence can be leveraged to build credibility, shape market perceptions, and drive business results. AI can partner in this process, helping streamline tasks, generate insights, and optimize analyst relations strategy.

CHAPTER 12

Customer Marketing

Customer Story Development

These are the heart and soul of customer marketing. They're how we showcase the real-world impact of our product, using customer successes to build trust and credibility. But crafting compelling customer stories takes time and effort. We need to gather information, interview customers, and weave those details into a narrative that resonates.

Instead of starting from scratch, we can feed the AI customer quotes, interview transcripts, and even survey data, and it can help generate compelling narratives that showcase the value of our product.

Example prompt

<context>
We need a customer success story that highlights how our analytics tool helped a mid-sized enterprise reduce costs.
</context>
<task>
Draft a customer success story highlighting how our analytics tool reduced costs by 30% for a mid-sized enterprise. Focus

on the challenges they faced, the solution we provided, and the quantifiable results they achieved.
</task>
<format>
Provide a story with a clear narrative structure, including a beginning, middle, and end, and incorporate customer quotes to add authenticity.
</format>

To ensure the story is authentic and resonates with the target audience, upload interview transcripts or customer testimonials to the AI. This will help it capture the customer's voice and tell a story that's both informative and engaging.

The goal is to create customer stories that inspire and persuade. AI can develop compelling narratives that showcase the real-world impact of a product.

Reference Programs

These are a powerful way to leverage the voices of our happiest customers, turning them into advocates who can help attract new business. Yet building and managing a successful reference program can be a lot of work. We need to identify potential references, nurture relationships, and provide them with the resources they need to be effective advocates.

AI can refashion this process and build a structured reference program that delivers results.

Instead of manually reaching out to customers and tracking their engagement, AI can automate many of the tasks involved.

Example prompt

<context>
We need to build a customer reference program to support our sales and marketing efforts.
</context>

Create a framework for a customer reference program, including:
- Criteria for selecting potential references.
- Outreach templates for inviting customers to participate.
- Resources and training for reference customers.
- Rewards and recognition for participation.
- Success metrics and tracking mechanisms.
</task>
<format>
Provide a structured framework with clear guidelines and actionable steps.
</format>

And here's an actionable step. Use AI to draft those customer outreach emails. You can even refine the messaging based on feedback from sales and marketing teams.

The goal is to build a community of loyal customer advocates willing to share their positive experiences with your product. AI can streamline the building and management of a successful reference program.

Customer Spotlight Creation

Like mini case studies, customer spotlights provide a concise snapshot of a customer's success with a product. They're great for sharing on social media, featuring in blog posts, or including in sales enablement materials.

The fact is writing effective customer spotlights can be demanding. The essence of the customer's story needs to be captured concisely and engagingly.

AI can do this, ensuring they're informative, impactful, and tailored to a target audience.

Example prompt

<context>
We need a customer spotlight for a healthcare company that's using our cybersecurity platform.

</context>
<task>
Write a 150-word spotlight for a healthcare company using our cybersecurity platform. Focus on the key benefits they've experienced, such as improved data protection, reduced security risks, and streamlined compliance.
</task>
<format>
Provide a spotlight with a compelling headline, a concise summary of the customer's success, and a relevant quote from the customer.
</format>

Pro tip:

Tailor spotlights for different channels. One for a blog post might be more detailed and narrative-driven, while for social media shorter and more visually oriented. Iterate on the AI's output to create variations that fit each channel.

The goal is to showcase customer success in a way that's both informative and engaging. AI can generate concise and compelling customer spotlights to be used across various marketing channels.

Testimonial Formatting

Testimonials are like gold dust for marketers. A powerful one can build trust, credibility, and social proof. Unfortunately, raw customer quotes often need some polishing before they're ready for prime time.

AI can transform those raw quotes into polished testimonials that are concise, impactful, and aligned with brand messaging.

Example prompt

<context>
We have a customer quote that talks about the scalability of our platform.

</context>

<task>

Rephrase this customer quote into a formal testimonial emphasizing the scalability benefits.

</task>

<format>

Provide a polished testimonial with correct grammar, punctuation, and a professional tone.

</format>

To get the best results, provide the AI with some specific guidelines:

- **Tone:** Do you want the testimonial to be formal or informal? Enthusiastic or matter-of-fact?
- **Length:** How long should the testimonial be? A short and punchy quote or a more detailed statement?
- **Focus:** What key message do you want to emphasize?

The goal is to create testimonials that are persuasive and credible. AI can transform raw customer quotes into polished statements that resonate with our target audience.

Practice Exercises

- **Customer story development challenge**: Select a hypothetical B2B product and develop a comprehensive customer success story. Interview a fictional customer (create a detailed persona) to capture their challenges, implementation journey, and quantifiable results. Create multiple variations of the story tailored for different marketing channels, demonstrating how to adapt the narrative for blog posts, social media, and sales enablement materials.

- **Reference program design workshop**: Design a complete customer reference program for a technology company. Develop a structured framework that includes detailed criteria for selecting reference customers, create outreach and communication templates, design a rewards system, and establish metrics for tracking the program's success. Consider how to leverage AI to streamline different aspects of the reference program management.

- **Customer testimonial and spotlight creation exercise**: Take a raw customer quote about a product's impact and transform it into multiple polished marketing assets. Create a formal testimonial, a social media-friendly spotlight, and a detailed case study excerpt. Demonstrate the ability to maintain the customer's authentic voice while aligning the content with specific marketing objectives and brand messaging.

Summary

- **AI-powered customer story creation**: AI can assist in generating compelling customer stories by leveraging existing customer quotes, interview transcripts, and survey data.
- **Streamlined reference program management**: AI can automate various tasks involved in building and managing a customer reference program, such as identifying potential references and drafting outreach templates.
- **Efficient customer spotlight and testimonial generation**: AI can help create concise and impactful customer spotlights and testimonials, tailored to specific audiences and channels.

CHAPTER 13

Analytics and Research

Metrics Analysis Framework

Data is the lifeblood of modern marketing. We're constantly tracking metrics, analyzing campaigns, and measuring ROI. But making sense of all that data can be overwhelming. A framework is needed, a structured approach to analyze metrics and extract actionable insights.

With AI, instead of manually crunching numbers and creating spreadsheets, AI develops frameworks for tracking and analyzing key product marketing metrics, such as lead generation, conversion rates, and campaign performance.

Example prompt

<context>
We need to track the ROI of our marketing campaigns.
</context>
<task>
Create a framework for tracking campaign ROI, including metrics like:
• Cost per lead (CPL).
• Customer acquisition cost (CAC).

- Conversion rates at each stage of the funnel.
- Revenue generated by campaign.
- Customer lifetime value (CLTV).

</task>
<format>
Provide a framework with clear definitions of each metric, suggested tracking mechanisms, and reporting templates.
</format>

Pro tip:

Use AI to generate periodic performance summaries. This can help identify trends, spot areas for improvement, and keep marketing efforts on track.

The goal is to turn data into actionable insights. AI can provide a structured framework for analyzing metrics and generating reports that highlight key trends and opportunities.

Survey Design

Surveys are a goldmine of customer insights, providing valuable feedback on products, messaging, and overall market positioning. Designing them to yield actionable results can be complex. We need to ask the right questions, phrase them clearly, and ensure we're covering all the key topics.

Instead of relying on gut feeling or outdated templates, use AI to craft survey questions that are clear, concise, and unbiased.

Example prompt

<context>
We need to assess customer satisfaction with our cybersecurity platform.
</context>
<task>

Draft 10 survey questions to assess customer satisfaction with our cybersecurity platform. Focus on areas like:
- Ease of use and deployment.
- Effectiveness of security features.
- Quality of customer support.
- Overall value and ROI.

</task>

<format>

Provide a mix of question types (e.g., multiple choice, rating scales, open-ended) and ensure the questions are phrased in a neutral and unbiased way.

</format>

And here's an actionable step. Use AI to iterate on the phrasing of questions. This can help ensure clarity, avoid ambiguity, and eliminate any potential bias that might skew results.

The goal is to gather accurate and insightful customer feedback. AI can craft effective survey questions that yield reliable data and actionable insights.

Survey Analysis

This is how we stay ahead of the curve, understand customers, and identify new opportunities. The task is to sift through mountains of research reports, competitive analyses, and survey results, which is daunting.

But using AI as a time-saver means not having to manually summarize key findings and extract insights. AI can synthesize information from multiple research sources and present it in a clear, concise, and actionable way.

Example prompt

<context>

We've conducted a customer survey to gather feedback on our new product.

</context>

<task>

Summarize the key findings from these uploaded customer survey responses into actionable insights. Focus on:

- Customer satisfaction with key features.
- Areas for improvement.
- Pricing and packaging preferences.
- Overall sentiment and recommendations.

</task>

<format>

Provide a summary with key findings, supporting data, and actionable recommendations.

</format>

To make this even more powerful, upload the raw survey data or even entire whitepapers to the AI. This will allow it to perform a deeper analysis and generate more comprehensive insights.

The goal is to turn research data into actionable strategies. AI can quickly synthesize information from multiple sources and present it in a way that's easy to understand and act upon.

Google Analytics Data Analysis

The go-to tool for understanding website traffic, user behavior, and campaign performance. But the sheer volume of data can be overwhelming. There needs to be a way to quickly identify key trends, spot areas for improvement, and translate those insights into action.

Instead of manually sifting through reports and dashboards, use AI to interpret and summarize Google Analytics data and provide actionable marketing insights.

Here's how it works:

<context>

We need to improve our website traffic and conversion rates.

</context>

<task>

Analyze this Google Analytics data and provide recommendations to improve website traffic and conversion rates. Focus on:

- Identifying top-performing pages and content.
- Analyzing user behavior (bounce rates, time on page, etc.).
- Identifying key traffic sources (organic, paid, social, etc.).
- Suggesting improvements to website design and navigation.

</task>
<format>
Provide a summary with key findings, supporting data, and actionable recommendations.
</format>

To make this even more powerful, upload CSV exports or dashboards from Google Analytics to the AI. This will allow it to perform a deeper analysis and generate more specific insights.

Here's an actionable step. Identify specific questions to answer and prompt the AI for targeted recommendations. For example:

- "How can we reduce the bounce rate on our blog page?"
- "What keywords drive the most organic traffic to our pricing page?"

The goal is to turn website data into actionable strategies. AI can do this by quickly analyzing Google Analytics data and providing recommendations for improving website performance and driving conversions.

Performance Benchmarking

Benchmarking is how we measure ourselves against the competition, identify areas for improvement, and track our progress over time. Gathering and analyzing benchmarking data is a time-consuming process. AI can simplify it, then generate benchmarking reports that provide valuable insights.

Instead of manually collecting data and creating spreadsheets, AI can gather information from various sources, compare our performance to industry averages, and identify areas where we excel or fall short.

Example prompt

We need to understand how our solution's adoption rate compares to competitors.
</context>
<task>
Generate a benchmarking report comparing our solution's adoption rate with competitors in the enterprise segment. Include metrics such as:
- Number of customers.
- Market share.
- Customer growth rate.
- Customer churn rate.
</task>
<format>
Provide a report with clear visualizations (e.g., charts, graphs) and concise explanations of the key findings.
</format>

AI can also be used to benchmark other metrics, such as website traffic, conversion rates, customer satisfaction, and social media engagement.

The goal is to gain a clear understanding of performance relative to the competition and identify areas for improvement. AI can produce comprehensive benchmarking reports that provide valuable insights and guide our strategic decision-making.

Practice Exercises

- **Comprehensive marketing analytics framework development**: Develop a detailed metrics analysis framework specific to your company's product marketing efforts. Create a holistic tracking system for key performance indicators, including lead generation, conversion rates, customer acquisition costs, and customer lifetime value. Design a reporting template that provides actionable insights and demonstrates how different metrics interconnect to tell a comprehensive story of marketing performance.

- **Market research synthesis challenge**: Conduct an in-depth market research analysis focused on a product's target market. Synthesize data from customer surveys, industry reports, and competitive analyses to generate a comprehensive report. Create an executive summary that highlights key market trends, customer pain points, and strategic recommendations for positioning a product, showcasing the ability to transform complex research into clear, actionable marketing strategies.
- **Performance benchmarking and customer insights project**: Design a multistep research initiative that includes:
  - Developing an unbiased survey to assess customer satisfaction and product perception.
  - Creating a competitive benchmarking report comparing your product against key competitors.
  - Generating tailored executive summaries for different stakeholders (leadership, product team, sales team).

Summary

- **AI-powered metrics analysis**: AI can help PMMs develop frameworks for tracking and analyzing key product marketing metrics, such as lead generation, conversion rates, and campaign performance.
- **Streamlined market research synthesis**: AI can efficiently synthesize information from multiple research sources, providing PMMs with clear, concise, and actionable insights.
- **Data-driven decision-making with Google Analytics**: AI can interpret and summarize Google Analytics data, offering actionable recommendations for improving website traffic, conversion rates, and overall marketing effectiveness.

CHAPTER 14

Mastering the Art of PMM Prompting

As we enter the final chapter, I'd like to step back and reflect on what's been learned. I hope that I've shared, with some amount of detail, how PMMs can leverage AI to transform workflows—from market research and messaging to campaign execution. By integrating AI tools into workflows, PMMs can minimize the repetitive drudgery and double down on delivering higher-impact results—all while maintaining control over quality and strategic direction.

Throughout the guide, we've explored key areas, including:

- **AI and human collaboration frameworks**: Defining roles for AI and humans to optimize efficiency.
- **Prompt engineering**: Crafting precise prompts to unlock AI's full potential.
- **Market understanding**: Using AI for trend discovery, TAM sizing, and competitive analysis.
- **Positioning and messaging**: Aligning AI outputs with customer pain points and unique value propositions.
- **Content and communications**: Building scalable content strategies and whitepapers with AI support.

Remember, key lessons include:

- **AI as a copilot, not a replacement**:
  - o AI can handle research, data synthesis, and content drafting but requires human expertise for strategic refinement.
- **The role of thoughtful prompting**:
  - o Crafting detailed and specific prompts ensures that AI delivers results that align with objectives and resonate with an audience.
- **Human-AI collaboration for optimal outcomes**:
  - o Combining AI's speed and data processing capabilities with human intuition and judgment produces the best results.
- **The importance of quality control**:
  - o Implementing review processes prevents errors, maintains brand consistency, and ensures alignment with business goals.

Top Six Prompting Best Practices

To get the most out of AI, PMMs must adopt thoughtful and deliberate prompting techniques. Clear, well-structured prompts not only yield accurate results but also save time by reducing the need for extensive revisions. Here are six essential practices to ensure prompts that drive optimal outcomes.

1. **Be specific and provide clear context**:
   - o Vague instructions lead to irrelevant outputs. Specify the task, role, and desired format to guide AI toward meaningful results.
   - o **Example**: "As a customer success manager, draft an email template addressing [specific customer concern]."
2. **Use clear markers to structure prompts**:
   - o Tags like <instruction/> content </instruction> or #content# provide clarity, helping AI navigate complex tasks with ease.
   - o Structured prompts ensure logical and coherent outputs, even for detailed assignments.

3. **Start simple and build complexity gradually**:
   o Begin with basic prompts to test initial outputs, then add layers of detail as needed.
   o **Example**: Start with "Summarize this article" before requesting "Summarize with key insights for IT directors."

4. **Break down complex tasks into manageable chunks**:
   o For large projects, use step-by-step prompts to maintain focus and coherence. This approach, known as chain-of-thought prompting, prevents overwhelm and improves accuracy.
   o **Example**: "Step 1: Analyze market trends. Step 2: Summarize competitor positioning."

5. **Provide relevant context and use variables for reusability**:
   o Context-rich prompts result in tailored outputs. Use variables like {audience} and {product} in templates to streamline tasks across different scenarios.
   o **Example**: "Generate a sales pitch for {audience} highlighting {product}'s {key feature}."

6. **Use AI to simulate interviews, feedback, and conversations**:
   o Simulated interactions provide fresh insights into customer pain points, competitive gaps, or messaging efficacy.
   o **Example**: "Act as a product user sharing feedback on [specific feature]. What are their concerns and suggestions?"

By applying these practices, PMMs can generate more actionable, accurate, and context-driven AI outputs while refining their overall workflows.

Recent Developments in AI for PMMs

AI tools are rapidly evolving, offering PMMs advanced capabilities to tackle complex challenges and streamline workflows. These developments open new opportunities for refining messaging, enhancing market analysis, and driving GTM strategies.

Advancements in OpenAI's models (o1 and o3)

OpenAI's o1 and the anticipated o3 models introduce enhanced reasoning capabilities, including chain-of-thought logic and reflective problem-solving. These features are ideal for addressing nuanced challenges such as competitive analysis, pricing strategies, and GTM planning.

Best *practices for complex prompts:*

- Use multi-step prompts for tasks requiring deeper analysis, such as responding to analyst queries or refining competitive positioning.
- Simulate scenarios to explore GTM tactics, customer segmentation, or product differentiation.

NotebookLM: Google's rising star

NotebookLM has quickly become a valuable tool for PMMs, offering robust capabilities for synthesizing information and creating actionable outputs.

Key applications

- Producing concise and tailored briefing guides.
- Capturing and refining personas with input from multiple sources.
- Summarizing podcasts to extract actionable insights for thought leadership and content strategies.

AI agents in SaaS platforms: From Salesforce to Gong

SaaS applications like Salesforce and Gong now integrate AI agents that provide significant value for PMMs.

Gong's *AI capabilities*

- Use Gong transcripts to verify messaging, refine persona assumptions, and align sales conversations with marketing strategies.

- **Examples**:
    - "Analyze this transcript for recurring objections and suggest messaging updates."
    - "Identify key customer pain points and map them to our current persona profiles."
    - "Summarize feedback on [specific feature] and recommend refinements."

These tools, combined with the right prompting techniques, equip PMMs to work smarter, refine their strategies, and deliver greater impact.

Call to Action

As the role of AI in product marketing continues to expand, PMMs have an unprecedented opportunity to elevate their impact. To make the most of these tools and techniques, here's how to take action today.

- **Experiment and refine prompts**:
    - Apply the prompting techniques outlined in this guide to real-world scenarios.
    - Start small, test results, and iteratively improve prompts to achieve the most effective outputs.
- **Document and share insights**:
    - Build a personal library of prompt templates and workflows to streamline future tasks.
    - Share your best practices, lessons learned, and successful applications with peers to contribute to the PMM community's collective growth.
- **Leverage emerging tools and resources**:
    - Stay up to date with advancements in AI tools like OpenAI's models, NotebookLM, and SaaS-based AI agents.
    - Use these tools to tackle more complex challenges, from refining personas to creating data-driven GTM strategies.
- **Collaborate across teams**:
    - Integrate AI-driven insights into cross-functional

workflows with sales, product, and customer success teams.

- o Use AI outputs to align messaging, enhance competitive positioning, and support data-driven decision-making.
- **Keep learning and adapting:**
  - o AI capabilities will continue to evolve, and so should your approach.
  - o Invest time in exploring new tools, refining strategies, and staying informed about industry trends.

By embracing these steps, you'll unlock AI's full potential to enhance workflows, scale impact, and drive meaningful results for your organization. The time to act is now—start experimenting, share your success, and shape the future of AI-powered product marketing.

Visit TinyTechGuides.com

Appendix:
Alphabetic List of Prompts

Acknowledgments

As 2024 neared a close, I was separated from my full-time job as a director of product marketing managers due to a large RIF at the company. It was unexpected and a complete shock to the system but if you're in tech long enough, it's bound to happen. It wasn't my first time nor will it be my last. But, out of the shock, this TinyTechGuide was born. At first, I thought I'd breeze through writing this book with AI assistance but as with any endeavor, it's taken longer than expected. Alas, it is finally completed.

As always, I need to thank my family. First, my wonderful wife Erin for putting up with me and continuing to support TinyTechGuides. To Andy and Chris, thank you for occasionally glancing away from TikTok to see what I was working on. To Brady, you bark so much but it got me off the treadmill I walk on while writing.

A big thanks to Josipa Ćaran Šafradin, for the TinyTechGuide design system, and the seventh cover art. Another creative masterpiece indeed.

To Peter Letzelter-Smith for his expert editing and proofreading. The book is clearer, more readable, and easier to interpret because of you.

To Nick Jewell for early discussion on the concept, manuscript review, and business ideas.

Melissa Burroughs, with much gratitude, I thank you for your careful review of the manuscript as well as your endorsement.

Steve Archut, thank you for the competitive prompt, And Ben Kennedy, thanks for the generative AI persona profile fail we discussed based on your LinkedIn profile.

To Dave Gerhardt and Rich Mendis for their glowing endorsements of this book.

Remember, it's not the tech that's tiny, just the book!™

Ever onward!